Skills Worksheet

Directed Reading A

Section: Elements
ELEMENTS, THE SIMPLEST SUBSTANCES
Write the letter of the correct answer in the space provided.

_____ **1.** A pure substance that cannot be broken down into simpler substances by physical or chemical means is called what?
   a. a material
   b. a mixture
   c. an element
   d. a chemical

Only One Kind of Atom

_____ **2.** What is a substance in which all of the "building-block" particles are identical called?
   a. an element
   b. a pure substance
   c. a mineral
   d. a solution

_____ **3.** Which of the following are the building-block particles for elements?
   a. atoms
   b. electrons
   c. protons
   d. neutrons

CLASSIFYING ELEMENTS

_____ **4.** Which of the following is NOT a physical property of an element?
   a. reactivity
   b. hardness
   c. melting point
   d. density

_____ **5.** Why does a helium-filled balloon float up when you let go?
   a. Helium is more dense than air.
   b. Helium is less dense than air.
   c. Krypton is less dense than helium.
   d. Air is less dense than helium.

Identifying Elements by Their Properties

Match the correct description with the correct term. Write the letter in the space provided.

_____ **6.** is a characteristic chemical property of elements

_____ **7.** can be identified by its unique properties

_____ **8.** combines with oxygen to form rust

_____ **9.** has a melting point of 1,495°C

a. element

b. flammability

c. iron

d. cobalt

GROUPING ELEMENTS

Categories of Elements

Use the terms from the following list to complete the sentences below.

nonmetals   metals
elements   metalloids

10. All _____ are either metals, metalloids, or nonmetals.

11. Elements that are shiny and conduct heat and electric current

    are _____.

12. Elements that are poor conductors of heat are _____.

13. Elements with properties of both metals and nonmetals are

    called _____.

Categories Are Similar

Match the correct description with the correct term. Write the letter in the space provided.

_____ **14.** elements that are malleable

_____ **15.** traits that identify elements in a category

_____ **16.** another name for metalloids

_____ **17.** elements that are dull

a. characteristic properties

b. semimetals

c. metals

d. nonmetals

Match the correct description with the correct term. Write the letter in the space provided.

_____ **18.** iodine, sulfur, neon

_____ **19.** lead, copper, tin

_____ **20.** silicon, boron, antimony

a. nonmetals

b. metalloids

c. metals

Name _____ Class _____ Date _____

Directed Reading A

Section: Compounds

Write the letter of the correct answer in the space provided.

_____ **1.** Which of the following substances is a compound?
  a. oxygen
  b. magnesium
  c. water
  d. copper

COMPOUNDS: MADE OF ELEMENTS

_____ **2.** What kind of substance is composed of two or more elements that are chemically combined?
  a. an element
  b. a compound
  c. a mixture
  d. a particle

_____ **3.** Which of the following elements combine to form the compound citric acid?
  a. hydrogen and oxygen
  b. carbon and oxygen
  c. hydrogen, carbon, and oxygen
  d. sodium, hydrogen, carbon, and oxygen

Chemical Reactions Form Compounds

_____ **4.** Which of the following is the process by which substances change into new substances?
  a. a chemical reaction
  b. a chain reaction
  c. a physical reaction
  d. an atomic reaction

PROPERTIES OF COMPOUNDS

_____ **5.** Which of the following statements about compounds is true?
  a. All compounds react with acid.
  b. Each compound has its own physical properties.
  c. Compounds are used to identify elements.
  d. Compounds are similar to elements.

Properties: Compounds Versus Elements

_____ **6.** Why are we able to eat sodium and chlorine in a compound?
  a. Sodium reacts violently with calcium.
  b. Chlorine is table salt.
  c. The compound is harmless.
  d. Sodium is a metal.

The Ratio of Elements in a Compound

_____ **7.** How do elements join to form compounds?
  a. never in the same ratio
  b. in a specific mass ratio
  c. randomly
  d. in a 1:8 mass ratio

BREAKING DOWN COMPOUNDS

Use the terms from the following list to complete the sentences below.

    carbonic acid   chemical change   carbon dioxide

8. The compound that helps give some drinks "fizz" is called

  _____.

9. When you open a soft drink, carbonic acid breaks down into

  _____ and water.

10. The only way to break down compounds is through

  a(n) _____.

COMPOUNDS IN YOUR WORLD

Compounds in Industry

Write the letter of the correct answer in the space provided.

_____ **11.** Which of the following compounds is broken down to make
    aluminum?
  a. mercury oxide
  b. aluminum oxide
  c. aluminum chloride
  d. magnesium oxide

Compounds in Nature

_____ **12.** Which of the following can form compounds from nitrogen in the air?
   a. bacteria
   b. pea plants
   c. animals
   d. all plants

_____ **13.** What type of compound do plants and animals use to make proteins?
   a. sugar
   b. ammonia
   c. carbon dioxide
   d. nitrogen compounds

_____ **14.** What do plants use during photosynthesis to make carbohydrates?
   a. soil
   b. carbon dioxide
   c. carbon monoxide
   d. oxygen

Skills Worksheet

Directed Reading A

Section: Mixtures

PROPERTIES OF MIXTURES

Use the terms from the following list to complete the sentences below.

mixture   compound
physical   identity

1. A combination of substances that are not chemically combined is

    called a(n) _____.

2. Two or more materials that combine chemically form a(n)

    _____.

3. In a mixture, the _____ of the substances doesn't change.

4. Mixtures can be separated through _____ changes.

Separating Mixtures Through Physical Methods

Match the correct description with the correct term. Write the letter in the space provided.

_____ **5.** used to separate crude oil

_____ **6.** used to separate a mixture of aluminum and iron

_____ **7.** used to separate the parts of blood

_____ **8.** used to separate sulfur and table salt

a. distillation

b. centrifuge

c. filter

d. magnet

The Ratio of Components in a Mixture

Write the letter of the correct answer in the space provided.

_____ **9.** Which of the following affects the color of granite?
   a. ratio of minerals
   b. amount of mixture
   c. temperatures of mixture
   d. weight of minerals

SOLUTIONS

_____ **10.** Which of the following is NOT true of solutions?
    a. They contain a solute.
    b. They contain evenly mixed substances.
    c. They contain a solvent.
    d. They look like two substances.

_____ **11.** When substances separate and spread evenly throughout a mixture, what is the process called?
    a. solute
    b. dissolving
    c. chemical change
    d. solubility

_____ **12.** What is the substance that is dissolved in a solution called?
    a. solute
    b. solvent
    c. compound
    d. mixture

_____ **13.** In a solution, what is the substance in which something dissolves called?
    a. solute
    b. solvent
    c. compound
    d. mixture

Use the terms from the following list to complete the sentences below.

    solvent   particles   alloy
    soluble   solution

14. Salt is _____ in water because it dissolves in water.

15. In a solution of two gases, the substance that is present in the

largest amount is called the _____.

16. A solid solution of metals or nonmetals dissolved in metal is

called a(n) _____.

17. A solution contains many small _____.

18. The particles in a(n) _____ are so small that they don't scatter light.

CONCENTRATION OF SOLUTIONS

Write the letter of the correct answer in the space provided.

_____ **19.** What is a measure of the amount of solute dissolved in a given amount of solvent called?

    a. solution   **c.** mixture

    b. concentration   **d.** solvent

_____ **20.** What is one way of expressing concentration?

    a. in grams of solute per milliliter of solvent

    b. in grams of solute per millimeter of solvent

    c. in grams of solvent per milliliter of solute

    d. in grams of solvent per millimeter of solute

_____ **21.** How does a concentrated solution differ from a dilute solution if both solutions have the same volume?

    a. The concentrated solution has more solvent.

    b. The concentrated solution has less solvent.

    c. The concentrated solution has more solute.

    d. The concentrated solution has less solute.

Solubility

Use the graph below to answer questions 22 and 23. Write the letter of the correct answer in the space provided.

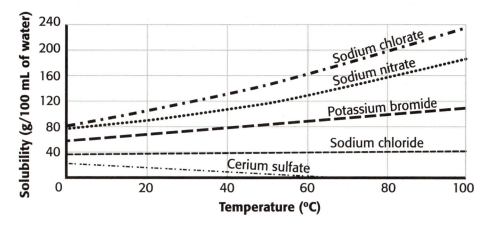

_____ **22.** Which solid is less soluble at higher temperatures than at lower temperatures?

    a. sodium chloride   **c.** sodium nitrate

    b. potassium bromide   **d.** cerium sulfate

_____ **23.** Which solid's solubility is least affected by temperature changes?

    a. cerium sulfate

    b. sodium nitrate

    c. potassium bromide

    d. sodium chloride

Use the terms from the following list to complete the sentences below.

temperature   solubility

24. The ability of one substance to dissolve in another at a given temperature and

pressure is called _____.

25. In a solution, the _____ usually affects the solubility.

Name _____ Class _____ Date _____

Directed Reading B

Section: Elements
ELEMENTS, THE SIMPLEST SUBSTANCES

1. A pure substance that cannot be separated into simpler substances by

   physical or chemical means is called a(n) _____.

2. A substance in which all of the "building-block" particles are identical

   is called a(n) _____ substance.

3. The building-block particles for elements are called _____.

CLASSIFYING ELEMENTS

4. The amount of an element present does not affect the element's

   _____.

5. Why does a helium-filled balloon float up when it is released?

   _____

   _____

Look at each property listed below. If it is a characteristic property of elements, write CP in the space provided. If it is not a characteristic property, write N.

_____ **6.** size

_____ **7.** melting point

_____ **8.** density

_____ **9.** shape

_____ **10.** mass

_____ **11.** volume

_____ **12.** color

_____ **13.** hardness

_____ **14.** flammability

_____ **15.** weight

_____ **16.** reactivity with acid

GROUPING ELEMENTS

17. What are two common properties that most terriers share?

_____

_____

18. All elements can be classified as metals, metalloids, or

_____.

19. An element that is shiny and that conducts heat and electricity well is called

a(n) _____.

20. An element that conducts heat and electricity poorly is called

a(n) _____.

21. Elements that have properties of both metals and nonmetals

are called _____.

Indicate whether the description applies to a metal, a nonmetal, or a metalloid. Write the correct letter in the space provided. Letters can be used more than once.

_____ **22.** are malleable

_____ **23.** are dull or shiny

_____ **24.** are poor conductors

_____ **25.** tend to be brittle and unmalleable as solids

_____ **26.** are almost always shiny

_____ **27.** are also called semimetals

_____ **28.** are almost always dull

_____ **29.** are somewhat ductile

_____ **30.** include boron, silicon, antimony

_____ **31.** include lead, tin, copper

_____ **32.** include sulfur, iodine, neon

a. metalloids

b. nonmetals

c. metals

Name _____ Class _____ Date _____

Directed Reading B

Section: Compounds

1. List three examples of compounds you encounter every day.

_____

_____

_____

COMPOUNDS: MADE OF ELEMENTS

2. When two or more elements are joined by chemical bonds to form a new pure substance, the new substance is called a(n) _____.

3. A compound is different from the _____ that make it up.

4. A(n) _____ is the process by which substances change into new substances.

PROPERTIES OF COMPOUNDS

_____ 5. Which of the following statements is true about the properties of compounds?
   a. A property of all compounds is to react with acid.
   b. Each compound has its own physical properties.
   c. Compounds cannot be identified by their chemical properties.
   d. A compound has the same properties as the elements that form it.

_____ 6. Which of the following is NOT true about compounds?
   a. Compounds are combinations of elements that join in specific ratios according to their masses.
   b. The mass ratio of a specific compound is always the same.
   c. Compounds are random combinations of elements.
   d. Different mass ratios mean different compounds.

7. Sodium and chlorine can be extremely dangerous in their elemental form. How is it possible that we can eat them in a compound?

_____

_____

_____

Match the correct description with the correct term. Write the letter in the space provided.

_____ **8.** a poisonous, greenish yellow gas

_____ **9.** table salt

_____ **10.** a soft, silvery white metal that reacts violently with water

a. sodium chloride

b. chlorine

c. sodium

BREAKING DOWN COMPOUNDS

11. What compound helps give carbonated beverages their "fizz"?

_____

12. Which elements make up the compound that helps give carbonated beverages their "fizz"?

_____

_____

_____

13. The only way to break down a compound is through a(n)

_____ change.

COMPOUNDS IN YOUR WORLD

14. Aluminum is produced by breaking down the compound

_____.

15. Plants use the compound _____ in photosynthesis to make carbohydrates.

Name _____ Class _____ Date _____

Directed Reading B

Section: Mixtures
PROPERTIES OF MIXTURES

1. A combination of two or more substances that are not chemically

combined is called a(n) _____.

2. When two or more materials combine chemically, they form a(n)

_____.

3. Each substance in a mixture keeps its _____.

4. How can you tell that a pizza is a mixture?

_____

_____

5. Mixtures can be separated through _____ changes.

Match each substance with the correct method of separation. Write the letter in the space provided. Each method may be used only once.

_____ **6.** a mixture of aluminum and iron

_____ **7.** crude oil

_____ **8.** parts of blood

_____ **9.** sulfur and salt

a. distillation

b. magnet

c. filter

d. centrifuge

10. Granite can be pink, gray, or black, depending on the

_____ of feldspar, mica, and quartz.

SOLUTIONS

_____ **11.** Which of the following is NOT true of solutions?
   a. They contain a dissolved substance called a solute.
   b. They are composed of two or more evenly distributed substances.
   c. They contain a substance called a solvent, in which another substance is dissolved.
   d. They appear to be more than one substance.

12. The process in which particles of substances separate and spread evenly

through a mixture is known as _____.

13. In a solution, the _____ is the substance that is dissolved,

and the _____ is the substance in which it is dissolved.

14. Salt is _____ in water because it dissolves in water.

15. When two gases or two liquids form a solution, the substance that is present

in the largest amount is the _____.

16. A solid solution of metals or nonmetals dissolved in metals is

a(n) _____.

17. What can particles in solution NOT do because they are so small?

_____

_____

CONCENTRATION OF SOLUTIONS

Use the graph below to answer questions 18 and 19. Write the letter of the correct answer in the space provided.

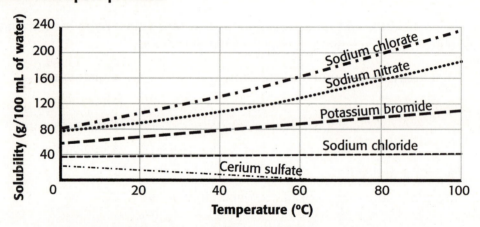

_____ **18.** Look at the graph above. Which solid is less soluble at higher
temperatures than at lower temperatures?
  a. sodium chloride
  b. sodium nitrate
  c. potassium bromide
  d. cerium sulfate

_____ **19.** Look at the graph above. Which compound's solubility is least affected
by temperature changes?
  a. sodium chloride
  b. sodium nitrate
  c. potassium bromide
  d. cerium sulfate

20. A measure of the amount of solute dissolved in a given amount of solvent is

called _____.

21. What is the difference between a dilute solution and a concentrated solution?

_____

_____

22. The ability of a solute to dissolve in a solvent at a certain temperature and

pressure is called _____.

Skills Worksheet

Vocabulary and Section Summary A

Elements

VOCABULARY

In your own words, write a definition of the following terms in the space provided.

1. element

_____

_____

2. pure substance

_____

_____

3. metal

_____

_____

4. nonmetal

_____

_____

5. metalloid

_____

_____

SECTION SUMMARY

Read the following section summary.

• A substance in which all of the particles are alike is a pure substance.

• An element is a pure substance that cannot be broken down into anything simpler by physical or chemical means.

• Each element has a unique set of physical and chemical properties.

• Elements are classified as metals, nonmetals, or metalloids, based on their properties.

Skills Worksheet

Vocabulary and Section Summary A

Compounds

VOCABULARY

In your own words, write a definition of the following term in the space provided.

1. compound

_____

_____

_____

SECTION SUMMARY

Read the following section summary.

- A compound is a pure substance composed of two or more elements.

- During a chemical reaction, the atoms of two or more elements react with each other to form molecules of compounds.

- Each compound has unique physical and chemical properties that differ from those of the elements that make up the compound.

- Compounds can be broken down into simpler substances only by chemical changes.

Name _____ Class _____ Date _____

Vocabulary and Section Summary A

Mixtures

VOCABULARY

In your own words, write a definition of the following terms in the space provided.

1. mixture

_____

_____

_____

2. solution

_____

_____

_____

3. solute

_____

_____

_____

4. solvent

_____

_____

_____

5. concentration

_____

_____

_____

6. solubility

_____

_____

_____

Vocabulary and Section Summary A *continued*

SECTION SUMMARY

Read the following section summary.

- A mixture is a combination of two or more substances, each of which keeps its own characteristics.

- Mixtures can be separated by physical means, such as filtration and evaporation.

- A solution is a mixture that appears to be a single substance but is composed of a solute dissolved in a solvent.

- Concentration is a measure of the amount of solute dissolved in a given amount of solvent.

- The solubility of a solute is the ability of the solute to dissolve in a solvent at a certain temperature.

Skills Worksheet

Vocabulary and Section Summary B

Elements

VOCABULARY

After you finish reading the section, try this puzzle! The underlined words below are missing all their vowels. Write the completed words in the spaces provided. Terms may be used more than once.

1. Each <u>LMNT</u> can be classified by a unique set of physical and chemical properties.

_____

2. Elements that have properties of both metals and nonmetals are called <u>MTLLDS</u>, or <u>SMMTLS</u>.

_____

3. The elements iron, nickel, cobalt, lead, tin, and copper are all <u>MTLS</u>.

_____

4. A(n) <u>PR SBSTNC</u> is a substance in which all the "building-block" particles, or atoms, are identical.

_____

5. <u>NNMTLS</u> are elements that are dull and poor conductors of heat and electric current.

_____

6. The elements boron, antimony, and silicon are all <u>MTLLDS</u>.

_____

7. A substance that cannot be broken down into simpler substances by physical or chemical means is known as a(n) <u>LMNT</u>.

_____

8. Elements that are malleable, ductile, shiny, and good conductors of heat and electricity are grouped into a category called the <u>MTLS</u>.

_____

9. The elements iodine, sulfur, and neon are all <u>NNMTLS</u>.

_____

SECTION SUMMARY

Read the following section summary.

- A substance in which all of the particles are alike is a pure substance.

- An element is a pure substance that cannot be broken down into anything simpler by physical or chemical means.

- Each element has a unique set of physical and chemical properties.

- Elements are classified as metals, nonmetals, or metalloids, based on their properties.

Skills Worksheet

Vocabulary and Section Summary B

Compounds

VOCABULARY

After you finish reading the section, try this puzzle! First, use the clues below to unscramble the letters, and write the correct term(s) in the space provided.

1. the simplest substance: METLEEN

   _ _ _ _ □ _ _

2. compounds formed by plants through photosynthesis and broken down by plants and plant-eating animals for energy: RTBYEAHCSADRO

   _ _ □ _ _ _ _ _ □ _ _ _ _

3. made up of atoms of two or more different elements joined by chemical bonds: DPCOONUM

   _ _ _ □ _ _ _ _

4. a sample of matter, either a single element or a single compound, that has definite chemical and physical properties: URPE ESCUNBATS

   □ _ _ _   _ _ _ _ □ _ _ _ □

5. compound formed when magnesium reacts with oxygen: SMEINGUAM DIOEX

   _ _ _ _ _ □ □ _ _   □ _ _ _ _

Now, unscramble the boxed letters to reveal a fact about compounds.

6. A compound has _ _ _ _ _ _ _ _ _ _ _ that are different from those of the elements that formed the compound.

SECTION SUMMARY

Read the following section summary.

- A compound is a pure substance composed of two or more elements.

- During a chemical reaction, the atoms of two or more elements react with each other to form molecules of compounds.

- Each compound has unique physical and chemical properties that differ from those of the elements that make up the compound.

- Compounds can be broken down into simpler substances only by chemical changes.

Name _____ Class _____ Date _____

Vocabulary and Section Summary B

Mixtures
VOCABULARY

After you finish reading the section, try this puzzle! Identify each term described by the clues. Then, find the terms in the word search puzzle. Words may appear forward, backward, horizontally, vertically, or diagonally.

_____ **1.** the ability of one substance to dissolve in another at a given temperature and pressure

_____ **2.** in a solution, the substance in which the solute dissolves

_____ **3.** can be expressed as grams of solute per milliliter of solvent (g/mL)

_____ **4.** a combination of two or more substances that are not chemically combined

_____ **5.** in a solution, the substance that is dissolved

_____ **6.** brass, salt water, and air, for example

N	R	B	V	Q	Y	D	N	D	F	Y	U	C	S
W	O	Q	P	M	K	O	Y	Z	P	T	M	A	O
D	Z	I	E	Z	I	X	B	Z	P	I	D	H	L
O	V	Y	T	T	Y	H	D	V	C	L	X	F	U
H	I	S	U	A	Z	Q	F	L	L	I	I	A	T
U	K	L	O	F	R	S	E	A	K	B	E	C	E
O	O	R	B	L	U	T	D	I	F	U	G	L	T
S	W	R	Z	P	V	N	N	C	W	L	X	E	P
M	I	X	T	U	R	E	P	E	A	O	L	U	U
H	C	H	F	O	S	T	N	E	C	S	S	N	S
G	R	C	Y	Q	K	O	Y	T	O	N	E	X	E
H	P	F	F	M	S	W	I	Y	D	R	O	C	K
S	B	S	O	R	W	M	V	G	G	C	W	C	S

SECTION SUMMARY

Read the following section summary.

- A mixture is a combination of two or more substances, each of which keeps its own characteristics.

- Mixtures can be separated by physical means, such as filtration and evaporation.

- A solution is a mixture that appears to be a single substance but is composed of a solute dissolved in a solvent.

- Concentration is a measure of the amount of solute dissolved in a given amount of solvent.

- The solubility of a solute is the ability of the solute to dissolve in a solvent at a certain temperature.

Skills Worksheet

Reinforcement

It's All Mixed Up

Complete this worksheet after you finish reading the section "Mixtures."

Label each figure below with the type of substance it BEST models: compound, solution, or element.

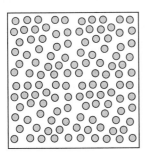

1

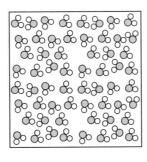

2

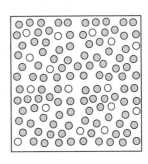

3

4. Why did you label the figures above as you did?

_____

_____

_____

_____

_____

_____

| Reinforcement *continued*

PROFESSOR JUMBLE'S CONFUSION

In her lab, Professor Jumble has two shelves labeled "Solutions" and "Compounds," respectively. Last night, the professor set one beaker of clear liquid on each of the two shelves. When the professor walked into her lab this morning, both beakers were on the same shelf, and she didn't know which was which. She tested each beaker, and the results are below.

5. Use the test results to help Professor Jumble unjumble the beakers, and write the identity of each liquid in the blanks.

Beaker A: _____

- Light passes right through.

- Particles do not separate in a centrifuge or a filter.

- Upon heating, the liquid evaporates, but no residue remains.

- The particles could not be separated by any other physical changes.

Beaker B: _____

- Light passes right through.

- Particles do not separate in a centrifuge or a filter.

- Upon heating, the liquid evaporates, and a crystal powder remains.

Skills Worksheet)

Critical Thinking

Jet Smart

You receive this letter from a top secret airplane manufacturer:

Agent X:

We were impressed by your work on our flying saucer project. Your help is now needed in the design of our newest stealth airplane, the FX-2000. We need your help with one simple but important matter—selecting the best metal for the plane's engines. Our team has narrowed the choices to two metals: titanium and platinum. Your mission is to gather facts about titanium and platinum, compare their properties, and recommend the better material. Report your answer within 24 hours.

You immediately turn to your reference books and study the properties of the two metals.

USEFUL TERM

corrosion wearing away gradually by rusting or the action of chemicals

Platinium

- a precious metal
- density: 21.4 g/cm^3
- resists corrosion
- melting point: 1,772°C
- weaker than steel

Titanium

- a metal
- density: 4.51 g/cm^3
- resists corrosion
- melting point: 1,675°C
- as strong as steel

MAKING COMPARISONS

1. How are platinum and titanium similar? How are they different?

_____

_____

_____

_____

_____

DEMONSTRATING REASONED JUDGMENT

2. Think about the extreme conditions that materials in the engine of a jet must endure. What properties would a metal in this engine need to have?

_____

_____

_____

_____

_____

3. Which material would you recommend? Explain your answer.

_____

_____

_____

_____

_____

PREDICTING CONSEQUENCES

4. Assume that the raw materials will be mined and sent directly to the manufacturing plant without being purified. Predict the possible consequences to the *FX-2000*'s performance. Explain your answer.

_____

_____

_____

_____

_____

Name _____ Class _____ Date _____

SciLinks Activity

MIXTURES

Go to www.scilinks.org. To find links related to mixtures, type in the keyword HY70974. Then, use the links to answer the following questions about mixtures.

Internet Resources

For a variety of links related to this chapter, go to www.scilinks.org

Topic: Mixtures
SciLinks code: HY70974

1. What are three key words or phrases related to solutions?

_____

_____

_____

2. What are three examples of solutions?

_____

_____

_____

3. Use your own words to explain a "homogeneous" mixture.

_____

_____

_____

4. Using four of the words or phrases you have recorded, create a concept map on a separate sheet of paper about solutions.

Name _____ Class _____ Date _____

Section Review

Elements
USING VOCABULARY

1. Use *element* and *pure substance* in the same sentence.

_____

_____

_____

UNDERSTANDING CONCEPTS

2. Classifying Compare the properties of metals and nonmetals.

_____

_____

_____

CRITICAL THINKING

3. Applying Concepts From which category of elements would you choose to make a container that would not shatter if dropped? Explain your answer.

_____

_____

_____

_____

4. Making Inferences List four possible properties of a substance classified as a metalloid. Can your list be used to classify an unknown substance as a metalloid? Explain your answer.

_____

_____

_____

_____

_____

| Section Review *continued*

MATH SKILLS

5. Making Calculations There are 8 elements that together make up 98.5% of Earth's crust: oxygen, 46.6%; aluminum, 8.1%; iron, 5.0%; calcium, 3.6%; sodium, 2.8%; potassium, 2.6%; magnesium, 2.1%; and silicon. What percentage of Earth's crust is silicon? Show your work below.

CHALLENGE

6. Evaluating Assumptions Your friend tells you that a shiny element has to be a metal. Do you agree? Explain your reasoning.

_____

_____

_____

_____

_____

_____

_____

_____

_____

_____

_____

_____

_____

_____

_____

_____

Skills Worksheet

Section Review

Compounds
UNDERSTANDING CONCEPTS

1. Identifying What type of change is needed to break down a compound?

_____

_____

_____

INTERPRETING GRAPHICS

The chart below shows the composition of table sugar in percent by mass. Use the chart to answer the next two questions.

Composition of Table Sugar

51.5% Oxygen

42.1% Carbon

Hydrogen

2. Evaluating List the 3 elements that make up table sugar.

_____

_____

_____

_____

3. Analyzing What percentage by mass of table sugar is hydrogen?

_____

_____

_____

_____

CRITICAL THINKING

4. Applying Concepts Iron is a solid, gray metal. Oxygen is a colorless gas. When iron and oxygen chemically combine, rust is made. Rust has a reddish brown color. Why does rust differ from iron and oxygen?

_____

_____

_____

_____

_____

_____

_____

CHALLENGE

5. Analyzing Ideas A jar contains samples of the elements carbon and oxygen. Does the jar contain a compound? Explain your answer.

_____

_____

_____

_____

_____

_____

_____

_____

_____

_____

_____

_____

_____

Name _____ Class _____ Date _____

Section Review

Mixtures

USING VOCABULARY

Correct each statement by replacing the underlined term.

1. The <u>solvent</u> is the substance that is dissolved.

_____

2. A measure of the amount of solute dissolved in a solvent is <u>solubility</u>.

_____

UNDERSTANDING CONCEPTS

Interpreting Graphics

Use the graph below to answer the next two questions.

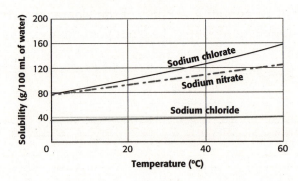

Solubility of Different Substances

3. Identifying At what temperature is 120 g of sodium nitrate soluble in 100 mL of water?

_____

4. Comparing At 60°C, how much more sodium chlorate than sodium chloride will dissolve in 100 mL of water?

_____

_____

5. Analyzing Identify the solute and solvent in a solution made of 15 mL of oxygen and 5 mL of helium.

_____

_____

_____

36 Elements, Compounds, and Mixtures

CRITICAL THINKING

6. Applying Concepts Soft drinks contain sugar and carbon dioxide. An open soda will lose carbonation. But, the soda will not become less sweet. Use the properties of the solutes to explain why.

_____

_____

_____

_____

_____

_____

7. Making Comparisons What are three ways that mixtures differ from compounds?

_____

_____

_____

_____

_____

_____

CHALLENGE

8. Applying Concepts Suggest a procedure by which to separate iron filings from sawdust. Explain why this procedure works.

_____

_____

_____

_____

_____

_____

_____

Skills Worksheet

Chapter Review

USING VOCABULARY

_____ 1. **Academic Vocabulary** In the sentence "The constituent elements of water are hydrogen and oxygen," what does the word *constituent* mean?
   a. empowered to elect
   b. component
   c. two
   d. only

Complete each of the following sentences by choosing the correct term from the word bank.

| compound | element | solution |
| solute | nonmetal | metal |

2. A(n) _____ has a definite ratio of components.

3. A(n) _____ is a pure substance that cannot be broken down into simpler substances by chemical means.

4. A(n) _____ is an element that is brittle and dull.

5. The _____ is the substance that dissolves to form a solution.

UNDERSTANDING CONCEPTS
Multiple Choice

_____ 6. Which of the following statements describes elements?
   a. All of the particles in the same element are different.
   b. Elements can be broken down into simpler substances.
   c. Elements have unique sets of properties.
   d. Elements cannot be joined in chemical reactions.

_____ 7. Which of the following best describes chicken noodle soup?
   a. element   **c.** compound
   b. mixture   **d.** solution

_____ 8. An element that conducts thermal energy well and is easily shaped is a
   a. metal.   **c.** nonmetal.
   b. metalloid.   **d.** None of the above

| Chapter Review *continued*

_____ **9.** Which of the following substances can be separated into simpler
substances only by chemical means?
a. sodium   **c.** water
b. salt water   **d.** gold

Short Answer

INTERPRETING GRAPHICS

The pie graphs below show the composition of citric acid and table sugar by
element (percentage by mass).

Use the pie graphs to answer the next three questions.

Composition of Citric Acid

58.3% Oxygen

Carbon

4.2% Hydrogen

Composition of Table Sugar

51.5% Oxygen

42.1% Carbon

Hydrogen

10. Analyzing What is the percentage by mass of carbon found in citric acid?

_____

_____

11. Identifying What is the difference between the percentage of hydrogen in
citric acid and the percentage of hydrogen in table sugar?

_____

_____

_____

12. Comparing Citric acid and table sugar are compounds. How can you tell from
the pie graphs that citric acid and table sugar are not the same compound?
Explain your reasoning.

_____

_____

_____

13. Comparing What is the difference between an element and a compound?

_____

_____

_____

_____

_____

14. Evaluating When nail polish is dissolved in acetone, which substance is the solute, and which is the solvent?

_____

_____

15. Evaluating Many gold rings are made out of 14-karat gold, which is an alloy of gold, silver, and copper. Is 14-karat gold a pure substance?

_____

_____

_____

WRITING SKILLS

16. Communicating Concepts On a separate sheet of paper, write an essay that could clearly explain to a third grade student the difference between elements, compounds, and mixtures. Your essay should have a thesis statement and include examples that support your ideas. Finally, make sure that your essay has a conclusion sentence.

❘ Chapter Review *continued*

CRITICAL THINKING

17. Concept Mapping Use the following terms to create a concept map: *matter, element, compound, mixture,* and *solution.*

18. Making Inferences A light green powder is heated in a test tube. A gas is given off, and the powder becomes a black solid. In which classification of matter does the green powder belong? Explain your reasoning.

_____

_____

_____

_____

_____

_____

Chapter Review *continued*

19. Applying Concepts Explain two properties of mixtures using a fruit salad as an example of a mixture.

_____

_____

_____

_____

_____

_____

_____

20. Forming Hypotheses Temperature affects the solubility of substances. Gases become less soluble as temperature increases. To keep the "fizz" in carbonated beverages after they have been opened, should you store them in a refrigerator or in a cabinet? Explain.

_____

_____

_____

_____

_____

_____

21. Analyzing Ideas Both carbon monoxide and carbon dioxide are made of carbon and oxygen, but they are not the same compound. Explain why these compounds differ from each other.

_____

_____

_____

_____

_____

_____

_____

22. Applying Concepts When hydrogen and oxygen react to form water, what happens to the atoms of the hydrogen and oxygen?

_____

_____

_____

_____

_____

_____

INTERPRETING GRAPHICS

Dr. Sol Vent did an experiment to find the solubility of a compound. The data below were collected using 100 mL samples of water.

Use the table below to answer the next two questions.

Temperature (°C)	10	25	40	60	95
Dissolved solute (g)	150	70	34	25	15

23. Forming Hypotheses Use a computer or graph paper to construct a graph of Dr. Vent's results. Examine the graph. To increase the solubility, would you increase or decrease the temperature? Explain.

_____

_____

_____

_____

_____

24. Predicting Consequences If 200 mL samples of water were used instead of 100 mL samples, how many grams of the compound would dissolve at 40°C?

_____

_____

_____

_____

_____

MATH SKILLS

25. Making Calculations What is the concentration of a solution prepared by dissolving 50 g of salt in 200 mL of water? Show your work below.

26. Making Calculations How many grams of sugar must be dissolved in 150 mL of water to make a solution that has a concentration of 0.6 g/mL? Show your work below.

CHALLENGE

27. Applying Concepts Describe a procedure that will separate a mixture of salt, finely ground pepper, and pebbles. Carefully consider the order in which you will perform each step. Explain why you chose the steps you did for each substance. How does knowing the properties of matter help you separate the substances in mixtures?

_____

_____

_____

_____

_____

_____

_____

_____

Assessment

Chapter Pretest

Teacher Notes and Answer Key

The Pretest questions are designed to help you determine the prior knowledge of your students. Some questions test whether students have mastered the background knowledge they need to understand the content you are about to teach. Other questions test your students' prior knowledge of the content you are about to teach. Use the Pretest with the Test Doctors and diagnostic teaching tips in these notes pages to help you tailor your instruction to your students' specific needs.

QUESTION NUMBER	CORRECT ANSWER	STANDARD
1	A	8.3.d
2	B	8.3.d
3	D	8.3.e
4	C	8.5.d
5	C	8.7.c
6	D	8.3.b
7	A	8.3.b
8	A	8.7.c
9	B	8.7.c
10	C	8.7.c

TEST DOCTOR

The following Pretest questions have been diagnosed by the Test Doctor. Find out what might be causing your students' "ailing" answers. Each Test Doctor is followed by a diagnostic teaching tip to help you address students' learning needs.

Question 1 *asks students to identify the characteristics of particles for the three states of matter.*

A **Correct.** The motion of particles changes depending on the state of matter.

B **Incorrect.** The shape of particles does not change with the state of matter.

C **Incorrect.** The size of particles does not change with the state of matter.

D **Incorrect.** The color of a substance does not change with the state of matter.

Diagnostic Teaching Tip: Students who have difficulty answering this question correctly might benefit from a review of the characteristics of particles in the three states of matter. Place students in groups of three. Have each group make a list of the characteristics of particles in the three states of matter. When the groups have finished, ask student volunteers to read points off their lists to make a master list for the class. Invite student volunteers to draw representations of particle behavior for each state of matter.

Question 2 *asks students to identify what happens to the motion of particles in a substance when the substance freezes.*

A **Incorrect.** The motion of particles becomes less random as a substance cools.

B **Correct.** The motion of particles slows down as a substance freezes.

C Incorrect. The motion of particles stops only at absolute zero.

D Incorrect. The motion of particles speeds up as a substance heats up.

Diagnostic Teaching Tip: Students who answer this question incorrectly might benefit from reviewing a graphic representation of particles in the three states of matter. Ask student volunteers to come up with sayings or other mnemonic devices to help them remember the varying particle motion, such as "When the Temperature Police say 'freeze,' you'd better slow down so you're barely moving." Interested students may want to look up the term *absolute zero.*

Question 3 *asks students to define the motion of molecules in a gas.*

A Incorrect. Molecules are not far apart and moving in the same direction in any of the states of matter.

B Incorrect. Molecules vibrate and are packed together closely in a solid.

C Incorrect. Molecules are close together but can move past one another in a liquid.

D Correct. Molecules move independently and collide frequently in a gas.

Diagnostic Teaching Tip: Students who have difficulty with this question might benefit from a visual representation of particle motion in the three states of matter. Take three jars of varying sizes; the smallest two should be practically the same size. Fill the smallest jar completely with marbles or some other uniform solid substance. Show students the filled container, and tell them this is how particles are found in a solid—packed closely together. Shake the jar, and tell students that particles vibrate in a solid. Then, pour the marbles in the medium-sized jar, and shake it. Show how the particles can move slightly now, but not too much. Finally, pour the marbles in the large jar, and shake it, showing how the particles can fly around randomly. Explain that this is how particles behave in a gas.

Question 4 *asks students to identify what happens when a substance changes from a liquid to a gas.*

A Incorrect. Changing from a liquid to a gas does not chemically change a substance.

B Incorrect. As a substance is heated, its particles gain energy.

C Correct. When a substance changes from a liquid to a gas, it changes form without a chemical change.

D Incorrect. The substance may not be water, which boils at 100°C.

Diagnostic Teaching Tip: Students who have trouble answering this question correctly might benefit from a review of what happens during a change of state. Review phase changes with students. Remind them how particles gain energy when a substance boils and that the substance changes its physical form without a chemical change.

Question 5 *asks students to define the properties used to identify a substance.*

A Incorrect. Density can be used to identify a substance.

B Incorrect. Hardness can be used to identify a substance.

C Correct. Mass is never used to identify a substance.

D Incorrect. The melting point can be used to identify a substance.

Diagnostic Teaching Tip: For students who have difficulty with this question, review the physical properties used to identify a substance. Have students call out properties that can be used to identify a substance. Make a list as the properties are called out. Ask why mass is not used. (Two pieces of the same substance could have a different size and thus a different mass.)

Question 6 *asks students to define the similarities between properties of a compound and those of its original components.*

A Incorrect. A compound will have different properties, including a different density, from its original components.

B Incorrect. A compound will have different properties (mass is not a property) from its original components.

C Incorrect. A compound will have different properties, which may or may not be similar, from its original components.

D Correct. A compound will have different properties from its original components.

Diagnostic Teaching Tip: Students who have difficulty answering this question might benefit from a visual demonstration of a substance having different properties from its original components. Show students a clean iron nail and an iron nail with rust. Tell students that the iron in the nail combined with oxygen in the air to make rust. Have student volunteers call out a few properties each for iron, oxygen, and rust. Have them note how the properties for each differ from each other. Be sure to emphasize Section 2, "Compounds," in Chapter 5, "Elements, Compounds, and Mixtures." Students must know that compounds have properties that are different from the properties of their constituent elements in order to master standard 8.3.b.

Question 7 *asks students to tell how a compound is formed.*

A Correct. A compound is formed by chemically combining elements.

B Incorrect. A compound is formed by chemically combining elements.

C Incorrect. A compound is not formed by combining properties.

D Incorrect. A compound is not formed by combining properties.

Diagnostic Teaching Tip: Students who have trouble with this question might benefit from a real-world example of a compound being formed. Have students think of a space shuttle launch. Tell them the orange-colored external tank contains one tank of liquid hydrogen and one tank of liquid oxygen. When the two elements are burned in the main engines, water vapor is produced. Be sure to emphasize Section 2, "Compounds," in Chapter 5, "Elements, Compounds, and Mixtures." Students must know that a compound is made by a combination of two or more elements in order to master standard 8.3.b.

Question 8 *asks students to define the properties that identify an element.*

A Correct. An element is a substance that cannot be separated or broken down into simpler substances by chemical means.

B Incorrect. A compound can be broken down chemically.

C Incorrect. Mixtures are not chemically combined but can be separated by physical means.

D Incorrect. A solution is a type of mixture.

Diagnostic Teaching Tip: Students who have difficulty answering this questions might benefit from seeing a carrot chopped into little pieces. Tell students that you can continue to chop the carrot until it is too small to cut, but it would still be a carrot. The same thing happens with elements. Be sure to emphasize Section 1, "Elements," in Chapter 5, "Elements, Compounds, and Mixtures." Students must know that substances can be classified by their properties in order to master standard 8.7.c.

Question 9 *asks students to determine that a mixture can be separated into its components by physical means.*

A Incorrect. Compounds can only be separated by chemical means.

B Correct. Mixtures can be separated by physical means.

C Incorrect. A solute is the substance that is dissolved in a solution.

D Incorrect. A solvent is the substance in which a solute is dissolved.

Diagnostic Teaching Tip: If students have difficulty with this question, create a mixture of small iron pieces (nails or bolts) and large aluminum pieces (aluminum cans). Separate them using a magnet. Discuss other ways you might separate this mixture (size of "particles"; color of "particles"). Point out that all successful ways are based on physical properties. Be sure to emphasize Section 3, "Mixtures," in Chapter 5, "Elements, Compounds, and Mixtures." Students must be able to understand that substances can be classified by their properties and that mixtures are not chemically combined in order to master standard 8.7.c.

Question 10 *asks students to identify a solution.*

A Incorrect. Table salt is a compound.

B Incorrect. Iodine is an element.

C Correct. Antifreeze is a solution (alcohol in water).

D Incorrect. Nickel is an element.

Diagnostic Teaching Tip: Students who have difficulty answering this question might benefit from a demonstration. Show students a glass of water. Then, pour some salt into the water and stir. It is now a solution of salt and water; a homogenous mixture that appears to be a single substance. The particles of salt are so small that they will not come out of solution. Discuss with students other types of solutions. Be sure to emphasize Section 3, "Mixtures," in Chapter 5, "Elements, Compounds, and Mixtures." Students must know that substances like salt water can be classified by their properties in order to master standard 8.7.c.

Name _____ Class _____ Date _____

Chapter Pretest

_____ **1.** States of matter depend on the
   A motion of particles.
   B shape of particles.
   C size of particles.
   D color of the substance.

_____ **2.** When a substance freezes, the motion of its particles
   A becomes more random.
   B slows down.
   C stops briefly.
   D speeds up.

_____ **3.** Which of the following best describes the molecules in a gas?
   A The molecules move in the same direction and are far apart.
   B The molecules vibrate and are packed together closely.
   C The molecules are close together but can move past one another.
   D The molecules move independently and collide frequently.

_____ **4.** What happens to a substance as it changes from liquid to gas?
   A The particles separate and combine, forming a new substance.
   B The particles in the substance lose energy.
   C The substance changes form without a chemical change.
   D The substance will remain at 100°C.

_____ **5.** Each of the following can be used to identify a substance
   EXCEPT for its
   A density.
   B hardness.
   C mass.
   D melting point.

_____ **6.** A compound will have
   A the same properties as its original components but a
      different density.
   B the same properties as its original components but a different mass.
   C similar properties as its original components.
   D different properties from its original components.

_____ **7.** A compound is formed by
   A chemically combining two or more different elements.
   B physically combining two or more different elements.
   C chemically combining two or more different properties.
   D physically combining two or more different properties.

Chapter Pretest *continued*

_____ **8.** If you are unable to break down a substance into a simpler substance either physically or chemically, the substance is
   A an element.
   B a compound.
   C a mixture.
   D a solution.

_____ **9.** You can separate iron shavings from aluminum shavings by using a magnet. This combination of iron and aluminum shavings is an example of
   A a compound.
   B a mixture.
   C a solute.
   D a solvent.

_____ **10.** Which of the following is a solution?
   A table salt
   B iodine
   C antifreeze
   D nickel

Assessment

Section Quiz

Section: Elements

Match the correct definition with the correct term. Write the letter in the space provided.

_____ **1.** a pure substance that cannot be separated into simpler substances by physical or chemical means

_____ **2.** a sample of matter, either a single element or a single compound, that has definite chemical and physical properties

_____ **3.** an element that is shiny and conducts heat and electricity well

_____ **4.** an element that conducts heat and electricity poorly and is dull in appearance

_____ **5.** an element that has the properties of both metals and nonmetals

a. metal

b. element

c. metalloid

d. nonmetal

e. pure substance

Write the letter of the correct answer in the space provided.

_____ **6.** Boiling point, melting point, and density are some of an element's
    a. nonreactive properties.
    b. physical properties.
    c. chemical properties.
    d. pure properties.

_____ **7.** A property of an element that does not depend on the amount of the element is called a(n)
    a. electromagnetic property.
    b. finite property.
    c. unique property.
    d. characteristic property.

_____ **8.** An element's ability to react with oxygen is an example of a
    a. pure substance.   **c.** chemical property.
    b. physical property.   **d.** melting point.

_____ **9.** An element is a pure substance in which there are how many kinds of atoms?
    a. two kinds of atoms   **c.** three kinds of atoms
    b. four kinds of atoms   **d.** one kind of atom

Name _____ Class _____ Date _____

Section Quiz

Section: Compounds

Write the letter of the correct answer in the space provided.

_____ **1.** When two or more elements join together chemically,
  a. a compound is formed.
  b. a mixture is formed.
  c. a substance that is the same as the elements is formed.
  d. the physical properties of the substances remain the same.

_____ **2.** The physical properties of compounds do NOT include
  a. melting point.
  b. density.
  c. reaction to light.
  d. color.

_____ **3.** Which of the following will NOT break down compounds?
  a. heat
  b. electric current
  c. chemical change
  d. filtering

_____ **4.** How do elements join to form compounds?
  a. randomly
  b. in a specific mass ratio
  c. in a ratio of 1 to 8
  d. as the scientist plans it

_____ **5.** Compounds found in all living things include
  a. proteins.
  b. ammonia.
  c. mercury oxides.
  d. carbonic acids.

_____ **6.** How do the properties of a compound compare with the properties of the elements that make up the compound?
  a. Only the physical properties are the same.
  b. Only the chemical properties are the same.
  c. All the properties are identical.
  d. The properties are different.

_____ **7.** By what processes can compounds be broken down?
  a. physical changes
  b. chemical changes
  c. compound changes
  d. either physical or chemical changes

Name _____ Class _____ Date _____

Section Quiz

Section: Mixtures

Match the correct definition with the correct term. Write the letter in the space provided.

_____ **1.** combination of two or more substances that are not chemically combined

_____ **2.** homogeneous mixture throughout which two or more substances are uniformly dispersed

_____ **3.** substance that dissolves in a solvent

_____ **4.** substance in which a solute dissolves

_____ **5.** amount of a substance in a given quantity of a mixture, solution, or ore

_____ **6.** ability of one substance to dissolve in another at a given temperature and pressure

_____ **7.** process that separates and spreads particles of substances evenly throughout a mixture

_____ **8.** description of a solution containing a relatively low concentration of solute

_____ **9.** process that separates a mixture based on the boiling points of the components

_____ **10.** machine that separates mixtures by the densities of the components

a. centrifuge

b. solute

c. solvent

d. dilute

e. mixture

f. dissolving

g. distillation

h. solution

i. concentration

j. solubility

Assessment

Chapter Test A

Elements, Compounds, and Mixtures
MULTIPLE CHOICE

Write the letter of the correct answer in the space provided.

_____ **1.** What kind of pure substance forms when two elements chemically combine?
   a. an element
   b. a compound
   c. a mixture
   d. a solution

_____ **2.** Which of the following is the process in which particles of substances separate and spread evenly throughout a mixture?
   a. filtration
   b. dissolving
   c. concentration
   d. distillation

_____ **3.** How can a compound be broken down?
   a. by physical changes
   b. by chemical changes
   c. by crushing
   d. by cooling

_____ **4.** In which of the following are particles of two or more substances evenly mixed so they appear to be a single substance?
   a. a compound
   b. a mixture
   c. a solution
   d. an element

_____ **5.** Which of the following is a chemical property?
   a. density
   b. flammability
   c. melting point
   d. color

_____ **6.** Which of the following is true about elements?
   a. They are impure substances.
   b. They cannot be classified by their properties alone.
   c. They cannot be broken down into simpler substances.
   d. They have more than one kind of particle.

❙ Chapter Test A *continued*

_____ **7.** Which of the following is NOT true of compounds?
   a. They contain two or more elements.
   b. They form after a physical change.
   c. They have their own physical properties.
   d. They do not form randomly.

_____ **8.** What can be said about the properties of a compound?
   a. They are different from the properties of the elements that form the compound.
   b. They are identical to the properties of the elements that form the compound.
   c. They are not unique.
   d. They are formed after a physical reaction.

MATCHING

Match the correct description with the correct term. Write the letter in the space provided.

_____ **9.** pizza

_____ **10.** nugget of gold

_____ **11.** water

_____ **12.** salt water

   a. mixture
   b. solution
   c. element
   d. compound

Match the correct description with the correct term. Write the letter in the space provided.

_____ **13.** unique set of characteristics used to classify elements

_____ **14.** grams of solute per milliliter of solvent (g/mL)

_____ **15.** a solid solution of metals or nonmetals dissolved in metals

_____ **16.** an element that shares metal and nonmetal properties

_____ **17.** a homogeneous mixture that appears to be a single substance

_____ **18.** the substance in which a solute dissolves

_____ **19.** simplest substance

   a. alloy
   b. solution
   c. solvent
   d. metalloid
   e. properties
   f. concentration
   g. element

FILL-IN-THE-BLANK

Use the terms from the following list to complete the sentences below.

| ratio | concentration | nitrogen |
| distillation | nonmetals | dilute |

20. Solutions can be described as being concentrated or

_____.

21. A mixture of liquids can be separated by _____.

22. Bacteria make compounds from _____ in the air.

23. All _____ are poor conductors of heat and electric

current.

24. Elements join in a specific mass _____ to form a

compound.

25. The amount of a particular substance in a given quantity of a mixture,

solution, or ore is known as the _____.

Assessment

Chapter Test B

Elements, Compounds, and Mixtures
MULTIPLE CHOICE
Write the letter of the correct answer in the space provided.

_____ **1.** What is a pure substance made of two or more elements that are chemically combined called?
   a. a solution
   b. a compound
   c. a mixture
   d. an element

_____ **2.** If a spoonful of salt is mixed in a glass of water, what is the water called?
   a. a solute
   b. a solution
   c. a solvent
   d. an element

_____ **3.** What is a solid solution of a metal or nonmetal dissolved in a metal called?
   a. an element
   b. an alloy
   c. a pure substance
   d. a compound

_____ **4.** An element contains only one kind of
   a. particle.
   b. property.
   c. protein.
   d. raw material.

_____ **5.** What is formed when particles of two or more substances are distributed evenly among each other?
   a. a compound
   b. solubility
   c. a solution
   d. an element

_____ **6.** The flammability of a substance is
   a. a chemical property.
   b. related to the density.
   c. a physical property.
   d. changeable.

_____ **7.** How is a compound different from a mixture?
  a. Compounds have two or more components.
  b. Each substance in a compound loses its characteristic properties.
  c. Compounds are commonly found in nature.
  d. Solids, liquids, and gases can form compounds.

_____ **8.** The particles in a solution
  a. cannot scatter light.
  b. can settle out.
  c. are insoluble.
  d. can pass through a fine filter.

_____ **9.** When elements form mixtures, the elements
  a. keep their original properties.
  b. react to form a new substance with new properties.
  c. combine in a specific mass ratio.
  d. always change their physical state.

_____ **10.** Which of the following is NOT true of compounds?
  a. The unique set of properties of a compound differ from the properties of the elements that make up the compound.
  b. The particles are made of atoms of two or more elements that are chemically combined.
  c. Different samples of any compound have the same elements in the same proportion.
  d. They can be separated by physical methods.

_____ **11.** How are metalloids NOT similar to metals?
  a. They have some properties of nonmetals.
  b. Some are shiny, while others are dull.
  c. They are somewhat malleable and ductile.
  d. Some are good conductors of electric current.

_____ **12.** Which of the following statements is true about elements?
  a. They cannot be classified by their properties alone.
  b. They cannot be broken down into simpler substances.
  c. They are not pure substances.
  d. They have multiple particle types.

MATCHING

Match the correct description with the correct term. Write the letter in the space provided.

_____ **13.** is also known as table salt

_____ **14.** is used by plants during photosynthesis

_____ **15.** is a nonmetal

a. carbon dioxide

b. sulfur

c. sodium chloride

Match the correct description with the correct term. Write the letter in the space provided.

_____ **16.** an aluminum pie plate

_____ **17.** amount of solute dissolved in a solvent

_____ **18.** calcium carbonate

_____ **19.** potting soil

_____ **20.** instant hot chocolate that is placed in hot water

_____ **21.** a steel crowbar

a. mixture of solids

b. solute

c. element

d. concentration

e. compound

f. alloy

MULTIPLE CHOICE

Use the graph below to answer questions 22 and 23. Write the letter of the correct answer in the space provided.

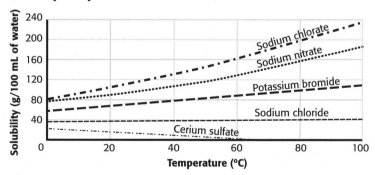

_____ **22.** Which solid is more soluble at lower temperatures than at higher temperatures?

    a. sodium chloride   **c.** potassium bromide

    b. sodium nitrate   **d.** cerium sulfate

_____ **23.** Which compound's solubility is least affected by changes in temperature?

    a. cerium sulfate   **c.** potassium bromide

    b. sodium nitrate   **d.** sodium chloride

Name _____ Class _____ Date _____

Chapter Test C

Elements, Compounds, and Mixtures
USING KEY TERMS

Use the terms from the following list to complete the sentences below. Each term may be used only once. Some terms may not be used.

solvent	solute	alloys
solution	mixture	pure substance
compound	metalloids	concentration

1. A pure substance made of two or more elements that are chemically

combined is called a(n) _____.

2. If a spoonful of salt is mixed in a glass of water, the salt is called

the _____.

3. Solid solutions of metals or nonmetals dissolved in metals

are called _____.

4. An element is a(n) _____ in which there is only one kind

of atom.

5. A measure of the amount of solute dissolved in a given amount of

solvent is referred to as _____.

6. Particles of two or more substances that are distributed evenly among

each other form a(n) _____.

UNDERSTANDING KEY IDEAS

Write the letter of the correct answer in the space provided.

_____ **7.** Which of the following is a chemical property?
  a. flammability   **c.** melting point
  b. density   **d.** ductility

_____ **8.** How is a mixture different from a compound?
  a. Mixtures have two or more components.
  b. Each substance in a mixture keeps its characteristic properties.
  c. Mixtures are commonly found in nature.
  d. Solids, liquids, and gases can form mixtures.

_____ **9.** The particles in a solution
  a. are insoluble.   **c.** can be filtered.
  b. can settle out.   **d.** cannot scatter light.

| Chapter Test C *continued*

_____ **10.** During what type of reaction do the atoms of two or more elements join together to form compounds?
   a. reaction with acid   **c.** chemical reaction
   b. physical reaction   **d.** chain reaction

_____ **11.** When materials combine to form a mixture, they
   a. keep their original properties.
   b. react to form a new substance with new properties.
   c. combine in a specific ratio.
   d. always change their physical state.

12. Why are compounds considered pure substances?

_____

_____

_____

13. How are metalloids different from metals?

_____

_____

_____

CRITICAL THINKING

14. Most fresh water in Saudi Arabia is produced by removing salt from seawater. One method involves distillation. Explain how seawater could be purified by distillation, and tell whether it is a chemical or physical process.

_____

_____

_____

15. Suppose you were given an unknown liquid and asked to determine if it was an element, a compound, or a mixture. What would you do? How would this help you find out what it was?

_____

_____

_____

_____

16. After testing three substances, a scientist has recorded data that identify some of the physical properties for each substance. Are these substances the same or different? Explain your answer.

Physical Property	Boiling point	Density	Color	Melting point	Hardness	Other
Substance 1		2.702g/cm³	silvery	660.25°C	2.75	nonmagnet
Substance 2	2,750°C	7.874g/cm³	silvery white		4	
Substance 3	2,467°C			660.25°C	2.75	nonmagnet

_____

_____

_____

INTERPRETING GRAPHICS

Use the graph to answer questions 17 and 18.

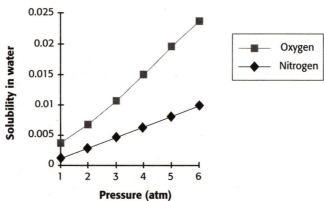

**Solubility in Water Versus Pressure
for Two Gases at 25°C**

17. What is the relationship between pressure and the solubility of oxygen and nitrogen in water?

_____

_____

18. Which gas experiences a greater change in solubility per unit pressure? Explain your answer.

_____

_____

_____

| **Chapter Test C** *continued*

CONCEPT MAPPING

19. Use the following terms to complete the concept map below:

pure substances   physical or chemical   chemical
solutions   compounds

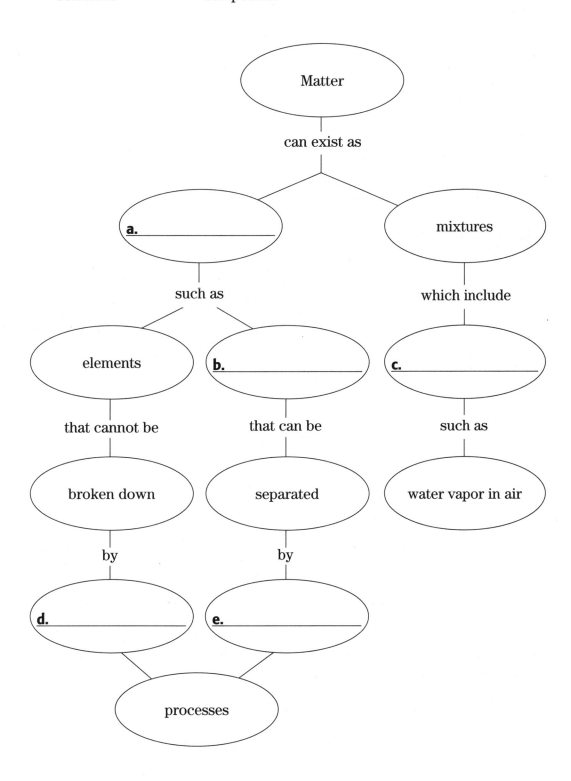

Assessment

Performance-Based Assessment

Teacher Notes

Terry Rakes
Elmwood Jr. High
Rogers, Arkansas

PURPOSE
Students observe the solubility of solids and gases in water at different temperatures.

TIME REQUIRED
One 45-minute class period. Students will need 25 minutes to form and test the hypothesis and 20 minutes to answer the analysis questions.

RATING
Easy ← 1 2 3 4 → Hard

Teacher Prep–1
Student Set-Up–2
Concept Level–3
Clean Up–1

ADVANCE PREPARATION
Equip each activity station with the necessary materials.

SAFETY INFORMATION
Students should not eat or drink anything in the laboratory. Do not use mercury thermometers. If a thermometer breaks, students should notify the teacher. Students should not heat glassware that is broken, chipped, or cracked. They should use tongs or heatproof gloves to handle heated glassware and other equipment. Clean up water spills immediately; spilled water is a slipping hazard. Never work with electricity near water; be sure the floor and all work surfaces are dry. Students should always wear heat-resistant gloves, goggles, and an apron when using a hot plate to protect their eyes and clothing. Never leave a hot plate unattended while it is turned on. Allow all equipment to cool before storing it. Students should tie back long hair, secure loose clothing, and remove loose jewelry.

TEACHING STRATEGIES
This activity works best in groups of 3–4 students. For this activity, "boiling" is the point at which students first see bubbles form on the bottom of the beaker. Show students how to correctly read a thermometer. In order to get an accurate reading, their eyes should be level with the mark. Ask students how the temperature could be read incorrectly if their eyes are above the level. *If your eyes are above the level, you will read a lower temperature than the actual temperature.* Ask students how the temperature could be read incorrectly if their eyes are below the level. *If your eyes are below the level, you will read a higher temperature than the actual temperature.*

Performance-Based Assessment *continued*

Evaluation Strategies

Use the following rubric to help evaluate student performance.

Rubric for Assessment

Possible points	Forming and testing hypothesis (30 points possible)
30–20	Successful completion of activity; safe and careful handling of materials and equipment; attention to detail; superior lab skills
19–10	Activity is generally complete; successful use of materials and equipment; sound knowledge of lab techniques; somewhat unfocused performance
9–1	Attempts to complete activity yield inadequate results; sloppy lab technique; no attention to detail; apparent lack of skill
	Analyzing results (40 points possible)
40–27	Superior analysis stated clearly and accurately; high level of detail; correct usage of scientific terminology
26–14	Accurate analysis; moderate level of detail; correct usage of scientific terminology
13–1	Erroneous, incomplete, or unclear analysis; incorrect use of scientific terminology
	Drawing conclusions (30 points possible)
30–20	Clear, detailed explanation shows good understanding of solubility of solids and gases in liquids at different temperatures; use of examples to support explanations
19–10	Adequate understanding of solubility of solids and gases in liquids at different temperatures with minor difficulty in expression
9–1	Poor understanding of solubility of solids and gases in liquids at different temperatures; explanation not relevant to the activity; factual errors

INQUIRY

Performance-Based Assessment

OBJECTIVE
You will investigate how solids and gases dissolve in water at different temperatures.

KNOW THE SCORE!
As you work through the activity, keep in mind that you will be earning a grade for the following:

- how well you form and test the hypothesis (30%)
- the quality of your analysis (40%)
- the clarity of your conclusions (30%)

Using Scientific Methods

ASK A QUESTION
Does the temperature of a liquid affect how much salt can dissolve in it?

MATERIALS AND EQUIPMENT

- beakers, 240 mL (2)
- gloves, heat-resistant
- hot plate
- paper towels
- salt, 240 mL

- spoon, 5 mL (1 tsp)
- stirring rod or spoon
- thermometer, alcohol
- thermometer holder
- water, cold, 240 mL (1 cup)

SAFETY INFORMATION

FORM A HYPOTHESIS

1. Form a hypothesis that explains whether salt will more easily dissolve in hot water or cold water.

TEST THE HYPOTHESIS

2. Pour 120 mL of cold water into each empty beaker.

3. Wearing goggles, heat-resistant gloves, and an apron, plug in and turn on the hot plate. Place one of the beakers on the hot plate. When the water begins to boil, place the thermometer in the water. Do not let the thermometer rest on the bottom of the beaker. Record the boiling point for water in degrees Celsius (°C).

4. Remove the beaker from the hot plate. Set it next to the other beaker.

5. To each beaker, add salt a spoonful at a time while stirring until no more salt dissolves. You will know when no more salt will dissolve when you see salt fall to the bottom no matter how much you stir. Record the amount of salt dissolved in the water. (Each spoonful is about 5 mL.)

Amount of Dissolved Salt

Cold water (mL)	Hot water (mL)

6. Place one of the beakers containing salt water on the hot plate. When the water begins to boil, place the thermometer in the water, and record the boiling point for salt water in degrees Celsius (°C).

_____

7. Turn off and unplug the hot plate.

ANALYZE THE RESULTS

8. Did the salt water boil at a lower temperature, the same temperature, or a higher temperature than the plain water?

_____

9. Did the cold water or the hot water dissolve more salt?

_____

10. Why did you see bubbles in the water when it heated up?

_____

_____

DRAW CONCLUSIONS

11. Compare how easily solids dissolve in hot water with how easily they dissolve in cold water. Write one sentence summarizing your observation.

_____

_____

_____

_____

Performance-Based Assessment *continued*

BIG IDEA QUESTION

12. In this experiment, you worked with water, table salt, and salt water. Classify each of these materials as an element, compound, or mixture. Then, explain why you classified them as such.

_____

_____

_____

_____

_____

_____

Assessment

Standards Assessment

Teacher Notes and Answer Key

To provide practice under more realistic testing conditions, give students 20 min to answer all of the questions in this assessment.

QUESTION NUMBER	CORRECT ANSWER	STANDARD
1	A	8.5.a (supporting)
2	C	8.3 (supporting)
3	B	8.3.b (supporting)
4	A	8.3 (supporting)
5	C	8.5 (supporting)
6	A	8.3.b (mastering)
7	A	8.7.c (mastering)
8	D	8.7.c (exceeding)
9	C	8.3.b (mastering)
10	D	8.3.b (exceeding)
11	D	8.7.c (exceeding)
12	A	6.3.c (mastering)
13	D	6.3.a (mastering)
14	D	8.3.e (mastering)
15	A	8.3.d (mastering)

TEST DOCTOR

The following Standards Assessment questions have been diagnosed by the Test Doctor. Find out what might be causing your students' "ailing" answers. Each Test Doctor is followed by a diagnostic teaching tip to help you address students' learning needs.

Question 1 *asks students to identify the meaning of* interact.

A **Correct.** *Interact* means to act upon one another.
B **Incorrect.** *Convert* means to change or alter.
C **Incorrect.** *Dissolve* means to pass into solution.
D **Incorrect.** *Transform* means to change.

> **Diagnostic Teaching Tip:** Students who have difficulty with this question might benefit from studying the roots of the word *interact*. Note that the prefix *inter-* means "between," so *interact* literally means "to act between."

Question 2 *asks students to define* distinct *in context.*

A **Incorrect.** This is not the correct definition for *distinct* in this sentence.
B **Incorrect.** In this sentence, *distinct* does not mean "notable."
C **Correct.** *Distinct* in this sentence means "separate and discrete."
D **Incorrect.** *Distinct* does not refer to visibility in this sentence.

Standards Assessment *continued*

Diagnostic Teaching Tip: Students who have difficulty answering this question correctly might benefit from writing sentences in which each definition of *distinct* is used. Invite students to share their sentences.

Question 3 *asks students to recognize the definition of* compound.

A Incorrect. A molecule might be made up of just one element.

B Correct. A compound is a chemical combination of two or more elements.

C Incorrect. A mixture is not chemically combined.

D Incorrect. A solution is not a chemical combination of elements.

Diagnostic Teaching Tip: Students who struggle to answer this type of question might benefit from illustrating the meanings of each answer choice by drawing the letters of each word as parts of a molecule, compound, mixture, or solution. For example, the word parts "com" and "pound" might be shown separately and then joined with chemical bonds.

Question 4 *asks students to define* structure *in context.*

A Correct. In this sentence, *structure* means arrangement.

B Incorrect. An atom's structure is not the same as its size.

C Incorrect. An atom's structure is just one of its properties.

D Incorrect. *Structure* is not defined as density.

Diagnostic Teaching Tip: Students who have difficulty with this question might benefit from describing the structure of several objects. For example, students might select several varied items, such as a baseball, a squirrel, or a school, and describe the structure of each one.

Question 5 *asks students to identify the correct form of the word* react.

A Incorrect. The verb *react* does not fit in this sentence.

B Incorrect. A reactor causes a reaction; it does not undergo one.

C Correct. Elements sometimes undergo a chemical reaction to form new substances.

D Incorrect. The adjective *reactive* is not the correct word for this sentence.

Diagnostic Teaching Tip: Students who have difficulty with this question might benefit from copying sentences from their textbook and taking out one key word from each. Have students swap sentences and decide what part of speech the missing word is, and then attempt to fill in the appropriate word.

Question 6 *asks students to demonstrate understanding of the properties of elements.*

A Correct. An element cannot be broken down into simpler substances by chemical means.

B Incorrect. Ions are formed when an atom gains or loses electrons.

C Incorrect. Bonds are interactions that hold atoms or ions together.

D Incorrect. Electrons are negatively charged particles in atoms.

Standards Assessment *continued*

Diagnostic Teaching Tip: Students who have difficulty answering this question should review the definitions and properties of elements, compounds, and mixtures. Have students make a table that compares these three groups of substances.

Question 7 *asks students to identify physical and chemical properties of elements.*

A Correct. The weight of a sample, plus the volume, would enable you to find out that element's density. Density is a characteristic property of elements, so this would be very helpful in identifying the element.

B Incorrect. The shape of a substance does not help determine what kind of element it is.

C Incorrect. The hardness of a substance would help narrow the choices down, but many elements have similar hardnesses, so this is not as helpful a characteristic as density.

D Incorrect. An element is not classified by where it is found.

Diagnostic Teaching Tip: Students who have difficulty answering this question correctly might benefit from an activity in which the students measure and compare the density of several elements. After students measure the density of known elements, give students a piece of an unknown element and ask them to identify the element.

Question 8 *asks students to identify the properties that help to identify an element.*

A Incorrect. These two substances have different densities, so they cannot be the same.

B Incorrect. These substances have different boiling points and densities, so they cannot be the same.

C Incorrect. These substances have the same boiling point, but different densities, so they cannot be the same.

D Correct. Substances A and D have the same boiling points and densities, so they are most likely the same.

Diagnostic Teaching Tip: Students who struggle to answer this question might benefit from brainstorming a list of properties and working with a group to classify which properties would most likely be the same for samples of the same substance.

Question 9 *asks students to demonstrate understanding of the properties of compounds and the elements from which they are formed.*

A Incorrect. A compound made of poisonous gases is not necessarily more poisonous than its elements.

B Incorrect. A compound made of poisonous gases might not be poisonous.

C Correct. The properties of a compound are not defined by the properties of its elements.

D Incorrect. The compound might be poisonous.

Diagnostic Teaching Tip: Students who struggle to answer this type of question might benefit from investigating the properties of elements and the compounds containing those elements. Make a list of elements and at least two compounds that contain each one. Have students investigate how the properties of the compounds differ from the elements they contain.

Question 10 *asks students to describe how compounds are broken down.*

A Incorrect. Dissolving and filtering may separate a mixture, but not a compound.

B Incorrect. Distillation is a physical change. It separates mixtures, but not compounds.

C Incorrect. A magnet can separate the metallic components of a mixture, but not a compound. Magnetic attraction is not a chemical change.

D Correct. A compound must be broken down with a chemical reaction, such as that which may be caused by an electrical current.

Diagnostic Teaching Tip: Students who struggle to answer this type of question might benefit from reviewing and comparing the processes of mixing and chemical reaction. Students could create flash cards for the terms *element, compound, mixture,* and *solution,* and write properties of each on the back, including how they are formed and how they break down.

Question 11 *asks students to identify a solution based on its structure.*

A Incorrect. Sample 1 is a compound, not a solution.

B Incorrect. The elements in Sample 2 are not evenly distributed, so it is a mixture.

C Incorrect. Sample 3 is a mixture, not a solution.

D Correct. Sample 4 is an evenly distributed mixture, so it is a solution.

Diagnostic Teaching Tip: Students who have difficulty with this question might benefit from practice identifying compounds, solutions, and mixtures from models. Have students illustrate the makeup of a compound, solution, or mixture, and then ask a partner to identify which of the three has been drawn.

Question 12 *asks students to identify the transfer of heat through a solid as conduction.*

A Correct. Heat is transferred through solids by conduction.

B Incorrect. Convection only occurs in liquids and gases.

C Incorrect. Refraction is not a process of transferring heat.

D Incorrect. Radiation is not the way that heat travels through a solid.

Diagnostic Teaching Tip: Students who have difficulty answering this question might benefit from reviewing the differences between conduction, convection, and radiation. Have students create a rhyme to help them remember the differences between these three processes.

Question 13 *asks students to demonstrate understanding of the process of heat flow.*

A Incorrect. Heat will flow between the two liquids.

B Incorrect. The warm water will cool down, but the cold water will also heat up.

C Incorrect. The cold water will heat up, but the warm water will also cool down.

D Correct. Heat will flow from the warm water to the cold water until they are both at the same temperature, which will be a temperature between those of the original cold and warm water samples.

Diagnostic Teaching Tip: Students who have difficulty with this question might benefit from conducting simple experiments that demonstrate how heat flows between substances. Have students mix ice cubes with liquids of different temperatures or place cold metal objects on warm surfaces and chart their observations.

Assessment

Standards Assessment

REVIEWING ACADEMIC VOCABULARY

_____ 1. Which of the following words means "to act upon one another"?
   A interact
   B convert
   C dissolve
   D transform

_____ 2. In the sentence "Argon has several properties distinct from other gases," what does *distinct* mean?
   A making a clear impression
   B notable
   C separate
   D easily seen

_____ 3. Which of the following words means "a substance formed by chemically combining two or more elements"?
   A molecule
   B compound
   C mixture
   D solution

_____ 4. In the sentence "Atoms of a certain element have a definite structure," which word is the closest in meaning to *structure*?
   A arrangement
   B size
   C property
   D density

_____ 5. Choose the word that best completes the sentence "Elements sometimes undergo a chemical _____ to form new substances."
   A react
   B reactor
   C reaction
   D reactive

REVIEWING CONCEPTS

_____ 6. Fundamental substances that cannot be broken down chemically into simpler substances are _____.
   A elements
   B ions
   C bonds
   D electrons

_____ **7.** Imagine that you were asked to classify four samples of equal and known volume, each of which was made up of a single element. Which factor would be most useful for identifying them?
   A mass
   B shape
   C hardness
   D original source

Properties of Substances

Substance	Mass (g)	Boiling Point (°C)	Density (kg/m³)
A	20	40	20
B	1,000	100	35
C	1,000	40	100
D	1,000	40	20

_____ **8.** Which two substances in the table are most likely the same?
   A substances A and C
   B substances B and D
   C substances C and D
   D substances A and D

_____ **9.** If two poisonous gases are combined chemically, which of the following will be true of the resulting compound?
   A The compound will be more poisonous than the gases.
   B The compound will be as poisonous as the gases.
   C The compound may or may not be poisonous.
   D The compound will not be poisonous.

_____ **10.** Which of the following processes can break down a compound?
   A dissolving and filtering
   B distilling at the boiling points of the compound's components
   C using a magnet to attract the compound's metallic components
   D applying an electric current

| Standards Assessment *continued*

The diagrams below represent the distribution of substances in four samples.

Sample 1
<AB> <AB> <AB>
   <AB> <AB> <AB>
<AB> <AB> <AB>

Sample 2
A B A A
B B A A
A B A A

Sample 3
<AB> A B B
A B A A
B <AB> A B

Sample 4
A B A B
B A B A
A B A B

_____ **11.** In the illustrations above, A and B are elements, and <AB> is a compound of elements A and B. Which illustration represents a solution of A and B?

A Sample 1
B Sample 2
C Sample 3
D Sample 4

REVIEWING PRIOR LEARNING

_____ **12.** When a solid copper block is heated at one end, the entire block is eventually heated. By what process is the heat transferred?

A conduction
B convection
C refraction
D radiation

_____ **13.** Which of the following is most likely to occur when warm water is mixed with cold water?

A The warm and cold water remain at their original temperatures.
B The mixed water soon reaches the same temperature as the cold water.
C The mixed water soon reaches the same temperature as the warm water.
D The mixed water soon reaches a temperature between the temperatures of the warm water and the cold water.

Explore Activity)

Classifying by Properties

Teacher Notes

In this activity, students will gain experience in classifying objects according to their properties. This activity will help students understand that substances can also be classified by their physical properties (covers standard 8.7.c).

MATERIALS

For each group

- An assortment of objects that can be classified by properties other than size and shape, such as color, metal/nonmetal, hard/soft, and solids/liquids (e.g., plastic items such as buttons, small wooden and metal objects, paper items, various liquids, etc.).

SAFETY CAUTION

Remind students to review all safety cautions and icons before beginning this activity.

Name _____ Class _____ Date _____

Classifying by Properties

In this activity, you will classify objects based on the properties that you can observe.

SAFETY INFORMATION

PROCEDURE

1. Look at the **assortment of objects.**
  • Classify the objects into two groups based on a physical property.
  • Do not use size or shape.
  • Record your classifications in the diagram provided on the following page.
  • List the objects in each group in the two top boxes in the diagram.
  • On the blank next to the row of boxes, write the physical property you used to classify the objects into groups. (This will be your answer key.)

2. Pick another physical property.
  • Classify the objects within each of the two groups based on this physical property.
  • Record the objects in the new groups in the boxes under the top two boxes.
  • Write the physical property you used in the space to the right of the boxes.

3. Repeat step 2 until each object is alone in a group.
  • Record each set of new groups in a new box on your diagram.
  • Remember to record which property you used each time.

4. Fold under the right side of your paper to hide your answer key. Switch diagrams with another student.
  • Pick out the property that the other student used in each step.
  • Write the property used in each step on a piece of paper.
  • Compare your guesses with the answer key of the student's paper.

Classifying by Properties *continued*

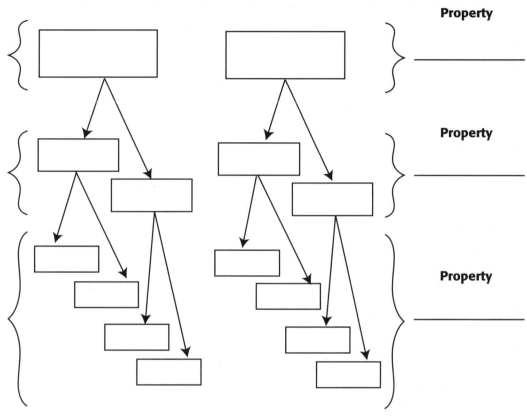

Property
_____

Property
_____

Property
_____

ANALYSIS

5. What properties did you use to classify your objects?

_____

_____

_____

_____

_____

6. Why do you think were you told not to use size and shape to classify the objects? (Hint: What if an ice cube were one of the objects?)

_____

_____

_____

_____

_____

Name _____ Class _____ Date _____

7. a. In what state (solid, liquid, or gas) is each of the objects? Write "S," "L," or "G" next to each object in your boxes.

_____

_____

_____

_____

_____

b. What can you learn about a substance's melting point and boiling point by observing the state of the substance at room temperature? (Hint: A liquid has already reached its melting point.)

_____

_____

_____

_____

_____

Name _____ Class _____ Date _____

Classifying by Properties

In this activity, you will classify substances based on observable properties.

SAFETY INFORMATION

PROCEDURE

1. Examine an **assortment of objects.** Classify the objects into two groups based on a physical property other than size or shape. Record which objects are in each group.

_____

_____

_____

_____

2. Choose another physical property. Based on this property, classify the objects within each group. Record your results.

_____

_____

_____

_____

3. Repeat step 2 until each object is alone in a group. Record the name of each group and object.

_____

_____

_____

_____

4. Switch lists with another student. Identify the property used in each step. Record your results.

_____

_____

_____

_____

Classifying by Properties *continued*

ANALYSIS

5. What properties did you use to classify your objects?

_____

_____

_____

_____

_____

_____

6. Why are an object's size and shape not reliable characteristics for classifying the object?

_____

_____

_____

_____

_____

_____

_____

7. In what state (solid, liquid, or gas) is each of these objects? What can you learn about a substance's melting point and boiling point by observing the state of the substance at room temperature?

_____

_____

_____

_____

_____

_____

_____

Name _____ Class _____ Date _____

Classifying by Properties

In this activity, you will classify substances based on observable properties.

SAFETY INFORMATION

PROCEDURE

1. Examine an **assortment of objects.** Classify the objects into two groups based on a physical property other than size or shape. Record the objects in each group.

_____

_____

_____

_____

2. Choose another physical property. Based on this property, classify the objects within each group. Record your results.

_____

_____

_____

_____

_____

_____

3. Repeat step 2 until each object is alone in a group. Record the name of each group and object.

_____

_____

_____

_____

_____

_____

_____

Classifying by Properties *continued*

4. Switch lists with another student. Identify the property used in each step. Record your results.

_____

_____

_____

_____

_____

ANALYSIS

5. What properties did you use to classify your objects, and why did you choose these properties?

_____

_____

_____

_____

_____

6. Why are an object's size and shape not reliable characteristics for classifying the object?

_____

_____

_____

_____

_____

_____

Classifying by Properties *continued*

7. Based on the state of each object at room temperature, what can you tell about its melting point and boiling point? How can you tell?

_____

_____

_____

_____

_____

_____

_____

_____

_____

_____

Quick Lab

Separating Elements

Teacher Notes

In this activity, students will learn that elements have properties (magnetism) that are not visible (covers standard 8.7.c). They will also learn that elements in a mixture can be separated by their properties.

MATERIALS

For each group

• magnet, bar

• nails, aluminum

• nails, iron

SAFETY CAUTION

Remind students to wear safety goggles and to handle sharp objects carefully.

Quick Lab

Separating Elements

SAFETY INFORMATION

PROCEDURE

1. Look at the sample of nails provided by your teacher.

2. Your sample has **aluminum nails** and **iron nails.**
   • Try to separate the two kinds of nails.
   • Make two piles (one for each kind of nail).

3. Now pass a **bar magnet** over each pile of nails. Record your results.

_____

_____

_____

4. Did you put the nails in the right piles? Explain.

_____

_____

_____

5. Think about a recycling plant.
   • Think about what you just did with the magnet.
   • Think about the properties of aluminum and iron.
   • How could a recycling plant use this information to separate aluminum and iron cans?

_____

_____

_____

_____

_____

_____

_____

Quick Lab   **DATASHEET B**

Separating Elements

SAFETY INFORMATION

PROCEDURE

1. Examine a sample of nails provided by your teacher.

2. Your sample has **aluminum nails** and **iron nails.** Try to separate the two kinds of nails. Group similar nails into piles.

3. Now pass a **bar magnet** over each pile of nails. Record your results.

_____

_____

_____

_____

4. Were you successful in completely separating the two types of nails? Explain.

_____

_____

_____

_____

_____

5. Based on your observations, explain how the properties of aluminum and iron could be used to separate cans in a recycling plant.

_____

_____

_____

_____

_____

Quick Lab

Separating Elements

SAFETY INFORMATION

PROCEDURE

1. Examine a sample of nails provided by your teacher.

2. Your sample has **aluminum nails** and **iron nails.** Try to separate the two kinds of nails. Group similar nails into piles.

3. Now pass a **bar magnet** over each pile of nails. Record your results.

_____

_____

_____

4. Were you successful in completely separating the two types of nails? Explain.

_____

_____

_____

5. Explain how the properties of aluminum and iron could be used to separate cans in a recycling plant.

_____

_____

_____

_____

_____

Quick Lab

Identifying Compounds

Teacher Notes

In this activity, students will learn how to use chemical reactivity to identify a compound (covers standards 8.5.a and 8.7.c). Exact amounts of baking soda are not needed (4 g is about 1 tsp). Add enough vinegar to just cover the solid in the cup.

MATERIALS

For each group

• baking soda, 4 g

• cup, clear plastic (2)

• sugar, powdered, 4 g

• vinegar, 10 mL

SAFETY CAUTION

Remind students to review all safety cautions and icons before beginning this activity.

Name _____ Class _____ Date _____

Identifying Compounds

SAFETY INFORMATION ⬦ ⬦ ⬦

PROCEDURE

1. Place **4 g of compound A** in a **clear plastic cup.**

2. Place **4 g of compound B** in a **second clear plastic cup.**

3. **a.** Look at compound A.

   • Note its color: _____
   • Note its texture (such as grainy, powdery, slippery, chunky):

   _____

   b. Look at compound B.

   • Note its color: _____
   • Note its texture (such as grainy, powdery, slippery, chunky):

   _____

4. Add **5 mL of vinegar** to each cup. Record your observations.

   • Cup with compound A: _____

   _____

   • Cup with compound B: _____

   _____

5. The two compounds are baking soda and powdered sugar. Baking soda reacts with vinegar. Powdered sugar does not react with vinegar. Fill in the blanks in the following sentences.

   I know compound A is _____

   because _____

   I know compound B is _____

   because _____

Quick Lab

DATASHEET B

Identifying Compounds

SAFETY INFORMATION ◆ ◆ ◆

PROCEDURE

1. Place **4 g of compound A** in a **clear plastic cup.**

2. Place **4 g of compound B** in a **second clear plastic cup.**

3. Observe the color and texture of each compound. Record your observations.

_____

_____

_____

_____

4. Add **5 mL of vinegar** to each cup. Record your observations.

_____

_____

_____

_____

5. Baking soda reacts with vinegar. Powdered sugar does not react with vinegar. Which compound is baking soda, and which compound is powdered sugar? Explain your answer.

_____

_____

_____

_____

_____

_____

Name _____ Class _____ Date _____

Identifying Compounds

SAFETY INFORMATION

PROCEDURE

1. Place **4 g of compound A** in a **clear plastic cup.**
2. Place **4 g of compound B** in a **second clear plastic cup.**
3. Observe the color and texture of each compound. Record your observations.

_____

_____

4. Add **5 mL of vinegar** to each cup. Record your observations.

_____

_____

5. The two compounds are baking soda and powdered sugar, which have very different reactions with vinegar. Research to find out how you would expect each compound to react with vinegar. Then identify compounds A and B. Explain your answer.

_____

_____

_____

_____

_____

_____

Quick Lab

Identifying Solutes by Solubility

Teacher Notes

In this activity, students will learn that solubility can be used to classify a substance (covers standard 8.7.c).

MATERIALS

For each group

- balance
- beaker, 50 mL (2)
- graduated cylinder
- salt, 25 g
- sugar, 25 g
- water, 20 mL

SAFETY CAUTION

Remind students to review all safety cautions and icons before beginning this activity.

Name _____ Class _____ Date _____

Identifying Solutes by Solubility

In water, sugar is almost eight times as soluble as table salt is. In this activity, you will use solubility to identify sugar and table salt.

SAFETY INFORMATION

PROCEDURES

1. Label one **50 mL beaker** as A.
   - Label a second 50 mL beaker as B.
   - Get a **graduated cylinder** for measuring.
   - Measure and pour **10 mL of water** into beaker A.
   - Measure and pour 10 mL of water into beaker B.

2. Use a **balance** to measure **2 g of compound A.**
   - Place it into beaker A.
   - Measure **2 g of compound B.**
   - Put it into beaker B.
   - Stir each mixture.

3. If both substances completely dissolve, repeat step 2.
   - You might have to complete step 2 more than once.

4. At some point, one of the compounds will no longer dissolve in the solution.

   - Which substance stopped dissolving (A or B)? _____
   - How much of each substance did you use?

   _____

   _____

5. Which substance is sugar and which is salt? How do you know?

   - Compound A is _____ because

   _____

   - Compound B is _____ because

   _____

   _____

Name _____ Class _____ Date _____

Identifying Solutes by Solubility

In water, sugar is almost eight times as soluble as table salt is. In this activity, you will use solubility to distinguish sugar from table salt.

SAFETY INFORMATION

PROCEDURE

1. Label one **50 mL beaker** as A and a second **50 mL beaker** as B. Using a **graduated cylinder,** measure and pour **10 mL of water** into each beaker.

2. Use a **balance** to measure **2 g of compound A** and place it into beaker A. Measure **2 g of compound B** and put it into beaker B. Stir each mixture.

3. If both substances completely dissolve, repeat step 2.

4. When one of the unknown substances no longer dissolves in the solution, record which substance stopped dissolving and how much of each substance was used.

_____

_____

_____

_____

_____

5. Identify which substance is sugar and which is salt.

_____

_____

_____

_____

_____

_____

_____

_____

Name _____ Class _____ Date _____

Identifying Solutes by Solubility

In water, sugar is almost eight times as soluble as table salt is. In this activity, you will use solubility to distinguish sugar from table salt.

SAFETY INFORMATION

PROCEDURES

1. Label one **50 mL beaker** as A and a second **50 mL beaker** as B. Using a **graduated cylinder**, measure and pour **10 mL of water** into each beaker.

2. Use a **balance** to measure **2 g of compound A** and place it into beaker A. Measure **2 g of compound B** and put it into beaker B. Stir each mixture.

3. If both substances completely dissolve, repeat step 2.

4. When one of the unknown substances no longer dissolves in the solution, record which substance stopped dissolving and how much of each substance was used.

_____

_____

5. Identify which solute is sugar and which is salt. How much more sugar do you think could be added to your sugar solution before the sugar stopped dissolving?

_____

_____

_____

_____

_____

_____

Flame Tests

Teacher Notes

In this activity, students will learn that tests can be performed to identify the elements that make up a compound (covers standards 8.3.b and 8.5.a). They will also learn that some substances (in this case, metals) can be classified by the distinct colors that they emit when heated (covers standard 8.7.c).

Jane Yuster
Hoover Elementary School
Redwood City, California1

TIME REQUIRED

One 45-minute class period

LAB RATINGS

Teacher Prep–3
Student Set-Up–2
Concept Level–4
Clean Up–3

Easy ◄——— 1   2   3   4 ———► Hard

MATERIALS

The materials listed on the student page are enough for each group of 2–3 students. The unknown solution should be clear. Use only dilute HCl (concentrations lower than 1.0 M). When diluting an acid, always add the acid to the water.

PREPARATION NOTES

Prepare solutions of KCl, CaC_{l2}, and NaCl in a concentration of 10 g/500 mL. Make enough of one of the solutions to serve as the "unknown." You will need 5 to 10 mL of each solution per group. Make the wire holder with Nichrome® wire or paper clips and corks or ice-cream sticks. Bend one end of the wire into a loop, similar to a bubble wand. Tape the other end of the wire to the stick, or insert it into the cork. If necessary, vinegar can be substituted for dilute hydrochloric acid in this experiment.

SAFETY CAUTION

Remind students to review all safety cautions and icons before beginning this activity. Students should touch only the wooden handle of the wire holder because the wire will become hot and could cause burns. Students should tie back long hair and secure loose clothing when they are working with the open flame. Students should be careful with the dilute HCl. If contact occurs, they should flush the skin immediately with water. In case of an acid spill, first dilute the spill with water. Then, mop up the spill with wet cloths or a wet mop while wearing disposable gloves.

Skills Practice Lab

Flame Tests

Fireworks produce fantastic combinations of color when they are ignited. The many colors are the results of burning many different compounds. Imagine that you are the head chemist for a fireworks company. The label has fallen off one box, and you must identify the unknown compound inside so that the fireworks may be used in the correct fireworks display. To identify the compound, you will use your knowledge that every compound has a unique set of properties.

OBJECTIVES

Observe flame colors emitted by various compounds.

Determine the composition of an unknown compound.

MATERIALS

- Bunsen burner
- chloride test solutions (4)
- hydrochloric acid, dilute, in a small beaker
- spark igniter

- tape, masking
- test tubes, small (4)
- test-tube rack
- water, distilled, in a small beaker
- wire and holder

SAFETY INFORMATION

Using Scientific Methods

ASK A QUESTION

1. How can you identify an unknown compound by heating it in a flame?

FORM A HYPOTHESIS

2. Write a hypothesis that is a possible answer to the question above. Explain your reasoning. (Hint: Think about what you see when you watch the flames in a bonfire.)

_____

_____

_____

_____

TEST THE HYPOTHESIS

3. Arrange the test tubes in the test-tube rack.
- Use masking tape to label each tube with one of the following names:
  calcium chloride
  potassium chloride
  sodium chloride
  unknown

4. Use the table below to record your data.

Test Results	
Compound	**Color of Flame**
Calcium chloride	
Potassium chloride	
Sodium chloride	
Unknown	

- Ask your teacher for the chemical solutions.
- **Caution:** Be very careful in handling all chemicals.
- Tell your teacher right away if you spill a chemical.

5. Light the burner.
- Clean the wire by dipping it into the dilute hydrochloric acid.
- Then dip it into distilled water.
- Hold the wooden handle of the wire.
- Heat the wire in the blue flame of the burner.
- Stop when the wire is glowing and it no longer colors the flame.
- **Caution:** Use extreme care around an open flame.

6. Dip the clean wire into the first test solution.
- Hold the wire at the tip of the inner cone of the burner flame.
- Record the color of the flame in the table.

7. Clean the wire by repeating step 5.
- Repeat steps 5 and 6 for the other solutions.

8. Follow your teacher's instructions for cleanup and disposal.

| Flame Tests *continued*

ANALYZE THE RESULTS

9. Identifying Patterns Is the flame color a test for the metal or for the chloride in each compound? Explain your answer. (Hint: Think about what is the same and different in the compounds.)

_____

_____

_____

_____

10. Analyzing Data What is the unknown solution? How do you know? (Hint: Did it burn like any of the other solutions?)

_____

_____

_____

_____

DRAW CONCLUSIONS

11. Evaluating Methods Why do you have to carefully clean the wire before testing each solution?

_____

_____

_____

12. Making Predictions Do you think the compound sodium fluoride will make the same color as sodium chloride in a flame test? Why or why not? (Hint: The flame color is based on the metal.)

_____

_____

_____

_____

_____

13. Interpreting Information Each compound you tested is made from chlorine. Chlorine is a poisonous gas at room temperature. Why is it safe to use these compounds without a gas mask? (Hint: Remember the difference between an element and a compound.)

_____

_____

_____

BIG IDEA QUESTION

14. Analyzing Methods You can use the flame test to see if sodium is one of the elements in a compound. But, you would not use the flame test on a pure metal to see if it is sodium. Explain why. (Hint: How is pure sodium different from sodium in a compound like sodium chloride?)

_____

_____

_____

_____

Skills Practice Lab

Flame Tests

Fireworks produce fantastic combinations of color when they are ignited. The many colors are the results of burning many different compounds. Imagine that you are the head chemist for a fireworks company. The label has fallen off one box, and you must identify the unknown compound inside so that the fireworks may be used in the correct fireworks display. To identify the compound, you will use your knowledge that every compound has a unique set of properties.

OBJECTIVES

Observe flame colors emitted by various compounds.

Determine the composition of an unknown compound.

MATERIALS

- Bunsen burner
- chloride test solutions (4)
- hydrochloric acid, dilute, in a small beaker
- spark igniter
- tape, masking
- test tubes, small (4)
- test-tube rack
- water, distilled, in a small beaker
- wire and holder

SAFETY INFORMATION

Using Scientific Methods

ASK A QUESTION

1. How can you identify an unknown compound by heating it in a flame?

FORM A HYPOTHESIS

2. Write a hypothesis that is a possible answer to the question above. Explain your reasoning.

_____

_____

TEST THE HYPOTHESIS

3. Arrange the test tubes in the test-tube rack. Use masking tape to label each tube with one of the following names: calcium chloride, potassium chloride, sodium chloride, and unknown.

Flame Tests *continued*

4. Copy the table below. Then, ask your teacher for your portions of the solutions. **Caution:** Be very careful in handling all chemicals. Tell your teacher immediately if you spill a chemical.

Test Results	
Compound	**Color of Flame**
Calcium chloride	
Potassium chloride	
Sodium chloride	
Unknown	

5. Light the burner. Clean the wire by dipping it into the dilute hydrochloric acid and then into distilled water. Holding the wooden handle, heat the wire in the blue flame of the burner until the wire is glowing and it no longer colors the flame. **Caution:** Use extreme care around an open flame.

6. Dip the clean wire into the first test solution. Hold the wire at the tip of the inner cone of the burner flame. Record in the table the color that the burning solution gives to the flame.

7. Clean the wire by repeating step 5. Then, repeat steps 5 and 6 for the other solutions.

8. Follow your teacher's instructions for cleanup and disposal.

ANALYZE THE RESULTS

9. Identifying Patterns Is the flame color a test for the metal or for the chloride in each compound? Explain your answer.

_____

_____

_____

_____

10. Analyzing Data What is the identity of your unknown solution? How do you know?

_____

_____

_____

_____

DRAW CONCLUSIONS

11. Evaluating Methods Why is it necessary to carefully clean the wire before testing each solution?

_____

_____

_____

_____

12. Making Predictions Would you expect the compound sodium fluoride to produce the same color as sodium chloride in a flame test? Why or why not?

_____

_____

_____

_____

13. Interpreting Information Each of the compounds that you tested is made from chlorine, which is a poisonous gas at room temperature. Why is it safe to use these compounds without a gas mask?

_____

_____

_____

_____

BIG IDEA QUESTION

14. Analyzing Methods You can use the flame test to determine if sodium is one of the elements in a compound. But you would not use the flame test on a pure metal to determine if the metal is sodium. Explain why.

_____

_____

_____

_____

_____

Skills Practice Lab

Flame Tests

Fireworks produce fantastic combinations of color when they are ignited. The many colors are the results of burning many different compounds. Imagine that you are the head chemist for a fireworks company. The label has fallen off one box, and you must identify the unknown compound inside so that the fireworks may be used in the correct fireworks display. To identify the compound, you will use your knowledge that every compound has a unique set of properties.

OBJECTIVES

Observe flame colors emitted by various compounds.

Determine the composition of an unknown compound.

MATERIALS

- Bunsen burner
- chloride test solutions (4)
- hydrochloric acid, dilute, in a small beaker
- spark igniter
- tape, masking
- test tubes, small (4)
- test-tube rack
- water, distilled, in a small beaker
- wire and holder

SAFETY INFORMATION

Using Scientific Methods

ASK A QUESTION

1. How can you identify an unknown compound by heating it in a flame?

FORM A HYPOTHESIS

2. Write a hypothesis that is a possible answer to the question above. Explain your reasoning.

_____

_____

_____

TEST THE HYPOTHESIS

3. Arrange the test tubes in the test-tube rack. Use masking tape to label each tube with one of the following names: calcium chloride, potassium chloride, sodium chloride, and unknown.

4. In the space below, create a table to record the flame colors of the four chloride test solutions. Ask your teacher for your portions of the solutions. **Caution:** Be very careful in handling all chemicals. Tell your teacher immediately if you spill a chemical.

5. Light the burner. Clean the wire by dipping it into the dilute hydrochloric acid and then into distilled water. Holding the wooden handle, heat the wire in the blue flame of the burner until the wire is glowing and it no longer colors the flame. **Caution:** Use extreme care around an open flame.

6. Dip the clean wire into the first test solution. Hold the wire at the tip of the inner cone of the burner flame. Record in the table the color given to the flame.

7. Clean the wire by repeating step 5. Then, repeat steps 5 and 6 for the other solutions.

8. Follow your teacher's instructions for cleanup and disposal.

Flame Tests *continued*

ANALYZE THE RESULTS

9. Identifying Patterns Which part of each compound is tested for by the flame test? Explain your answer.

_____

_____

_____

_____

10. Analyzing Data What is the identity of your unknown solution? How do you know?

_____

_____

_____

_____

_____

_____

DRAW CONCLUSIONS

11. Evaluating Methods Why is it necessary to carefully clean the wire throughout the lab?

_____

_____

_____

_____

12. Making Predictions Would you expect sodium fluoride to produce the same color as sodium chloride in a flame test? Why or why not?

_____

_____

_____

_____

_____

13. Interpreting Information Each of the compounds you tested is made from chlorine, which is a poisonous gas at room temperature. Why is it safe to use these compounds without a gas mask?

_____

_____

_____

_____

_____

_____

BIG IDEA QUESTION

14. Analyzing Methods Explain why you can use the flame test for sodium compounds, but you would not use the flame test on a pure metal to determine if it is sodium.

_____

_____

_____

_____

_____

_____

DATASHEET

Identifying Types of Parameters

Teacher Notes

In the Tutorial portion of this activity, students will learn the difference between variable and controlled parameters in a test. Then, students will plan an experiment using variable and controlled parameters (covers standard 8.9.c). In order to answer questions 4 and 5, students will need to conduct the experiment that they planned. Students should plan to dissolve Epsom salt in two or three samples of water of varying volumes (for example, 10 mL, 20 mL, and 30 mL).

MATERIALS

For each group

- balance
- beakers (2–3)
- Epsom salt (magnesium sulfate)
- graduated cylinder
- water

Name _____ Class _____ Date _____

DATASHEET

Identifying Types of Parameters

INVESTIGATION AND EXPERIMENTATION

8.9.c Distinguish between variable and controlled parameters in a test.

TUTORIAL

Many factors can affect the results of an experiment. So, when you plan an experiment, you must consider all of these factors or parameters. As part of your test plan, you will need to identify each parameter as controlled or variable.

1. Identify what you want to test. For example, you may want to test how temperature affects the solubility of a substance. The parameters that change during the experiment are the variable parameters. Your goal is to find out the relationship between the variable parameters.

2. List all of the other factors that could affect the results of your experiment. Examples of factors other than temperature that could affect the solubility of a substance are the amount of solute and the amount of solvent. To prevent these factors from affecting your test, you will need to keep their values constant. These factors that are kept constant are your controlled parameters.

3. When designing your experiment, you choose how much the values of a variable parameter will change between each measurement. For example, you might want to record the substance's solubility every 20°C, every 10°C, or even every 1°C.

4. The values of the controlled parameters will not change during your experiment.

5. After your experiment is complete, you could make a table or graph to determine the relationship between your variable parameters.

YOU TRY IT!

Plan an experiment to find out the relationship between the solubility of Epsom salt and the volume of water. You will do this by mixing increasing amounts of Epsom salt in water until no more will dissolve. Your parameters will be the amount of Epsom salt (in grams), the amount of water (in milliliters), and the temperature of the water (in degrees Celsius). If your plan is approved by your teacher, conduct the experiment.

1. Identifying What are your variable parameters? Why did you choose those parameters?

_____

_____

_____

2. Designing What is your controlled parameter? Why is having a controlled parameter during your experiment important?

_____

_____

_____

_____

3. Concluding What relationship did you find between your variable parameters?

_____

_____

_____

_____

4. Concluding Calculate the solubility of Epsom salt (g of Epsom salt/mL of water).

_____

_____

_____

_____

5. Designing If you needed to repeat this experiment, how would you change your design to improve your results? Explain your reasoning.

_____

_____

_____

_____

_____

_____

Answer Key

Directed Reading A

SECTION: ELEMENTS

1. C
2. B
3. A
4. A
5. B
6. B
7. A
8. C
9. D
10. elements
11. metals
12. nonmetals
13. metalloids
14. C
15. A
16. B
17. D
18. A
19. C
20. B

SECTION: COMPOUNDS

1. C
2. B
3. C
4. A
5. B
6. C
7. B
8. carbonic acid
9. carbon dioxide
10. chemical change
11. B
12. A
13. D
14. B

SECTION: MIXTURES

1. mixture
2. compound
3. identity
4. physical
5. A
6. D
7. B
8. C
9. A

10. D
11. B
12. A
13. B
14. soluble
15. solvent
16. alloy
17. particles
18. solution
19. B
20. A
21. C
22. D
23. D
24. solubility
25. temperature

Directed Reading B

SECTION: ELEMENTS

1. element
2. pure
3. atoms
4. characteristic properties
5. A helium-filled balloon will float up when released because helium is less dense than air.
6. N
7. CP
8. CP
9. N
10. N
11. N
12. CP
13. CP
14. CP
15. N
16. CP
17. Answers may vary. Sample answer: Terriers are small, and they have short hair.
18. nonmetals
19. metal
20. nonmetal
21. metalloids
22. C
23. A
24. B
25. B
26. C

27. A
28. B
29. A
30. A
31. C
32. B

SECTION: COMPOUNDS

1. Answers may vary. Sample answer: salt, water, and sugar
2. compound
3. elements
4. chemical reaction
5. B
6. C
7. Answers may vary. Sample answer: A compound has different properties from the elements that react to form it. Although sodium and chlorine are dangerous individually, they combine to form sodium chloride, a safe substance also known as table salt.
8. B
9. A
10. C
11. carbonic acid
12. carbon, oxygen, and hydrogen
13. chemical
14. aluminum oxide
15. carbon dioxide

SECTION: MIXTURES

1. mixture
2. compound
3. identity
4. Answers may vary. Sample answer: You can see each component in the pizza. Each component has the same chemical makeup as it did before the pizza was made.
5. physical
6. B
7. A
8. D
9. C
10. ratio
11. D
12. dissolving
13. solute; solvent (answers must be in this order)
14. soluble
15. solvent
16. alloy

17. Answers may vary. Sample answer: Particles in solution are so small that they can never settle out, cannot be removed or filtered out, and cannot scatter light.
18. D
19. A
20. concentration
21. Answers may vary. Sample answer: A dilute solution contains less solute than a concentrated solution does.
22. solubility

Vocabulary and Section Summary A

SECTION: ELEMENTS

1. element: a substance that cannot be separated or broken down into simpler substances by chemical means
2. pure substance: a sample of matter, either a single element or a single compound, that has definite chemical and physical properties
3. metal: an element that is shiny and that conducts heat and electricity well
4. nonmetal: an element that conducts heat and electricity poorly
5. metalloid: an element that has properties of both metals and nonmetals

SECTION: COMPOUNDS

1. compound: a substance made up of atoms of two or more different elements joined by chemical bonds

SECTION: MIXTURES

1. mixture: a combination of two or more substances that are not chemically combined
2. solution: a homogeneous mixture throughout which two or more substances are uniformly dispersed
3. solute: in a solution, the substance that dissolves in the solvent
4. solvent: in a solution, the substance in which the solute dissolves
5. concentration: the amount of a particular substance in a given quantity of a mixture, solution, or ore
6. solubility: the ability of one substance to dissolve in another at a given temperature and pressure

Vocabulary and Section Summary B

SECTION: ELEMENTS

1. element
2. metalloids, semimetals
3. metals
4. pure substance
5. nonmetals
6. metalloids
7. element
8. metals
9. nonmetals

SECTION: COMPOUNDS

1. element
2. carbohydrates
3. compound
4. pure substance
5. magnesium oxide
6. properties

SECTION: MIXTURES

1. solubility
2. solvent
3. concentration
4. mixture
5. solute
6. solution

N	R	B	V	Q	Y	D	N	D	F	Y	U	C	S
W	O	Q	P	M	K	O	Y	Z	P	T	M	A	O
D	Z	I	E	Z	I	X	B	Z	P	I	D	H	L
O	V	Y	T	T	Y	H	D	V	C	L	X	F	U
H	I	S	U	A	Z	Q	F	L	L	I	I	A	T
U	K	L	O	F	R	S	E	A	K	B	E	C	E
O	O	R	B	L	U	T	D	I	F	U	G	L	T
S	W	R	Z	P	V	N	N	C	W	L	X	E	P
M	I	X	T	U	R	E	P	E	A	O	L	U	U
H	C	H	F	O	S	T	N	E	C	S	S	N	S
G	R	C	Y	Q	K	O	Y	T	O	N	E	X	E
H	P	F	F	M	S	W	I	Y	D	R	O	C	K
S	B	S	O	R	W	M	V	G	G	C	W	C	S

Reinforcement

IT'S ALL MIXED UP

1. element
2. compound
3. solution

4. Answers may vary. Sample answer: In Figure 1, the particles are identical and part of the same substance, so it has to be an element. Figure 2 is a compound because the particles are identical but made of two different substances. Figure 3 is a mixture because it contains two different types of particles. Since it has a homogeneous mix of the two substances, it is a solution.
5. Beaker A is a compound, and Beaker B is a solution.

Critical Thinking

1. Answers may vary. Sample answer: Titanium and platinum are similar in that they are both metals that resist corrosion, and they have similar melting points. They differ in that platinum has a higher density and is a precious metal. Titanium is stronger than platinum.
2. Answers may vary. Sample answer: The metal would have to be lightweight to maximize the jet's speed, yet strong enough to withstand high pressure. It also should have a high melting point and should resist corrosion.
3. Answers may vary. Sample answer: I would choose titanium because it is stronger and lighter in weight than platinum. These properties would maximize durability and performance. It is probably also less expensive than platinum because it is not a precious metal.
4. Answers may vary. Sample answer: Most elements are found in nature as compounds. The properties of a compound are different from the properties of the elements that make up the compound. Possible consequences might be that the jet may be too heavy to fly, or the engine may overheat and melt.

SciLinks Activity

1. Answers may vary. Sample answer: dissolve, do not settle out, solute
2. Answers may vary. Sample answer: syrup, vinegar, salt water

3. Answers may vary. Sample answer: A homogeneous mixture is one in which the substances are evenly distributed throughout the solution.

4. Answers may vary but should show the relationships among four of the terms listed in response to questions 1 through 3 of this SciLinks Activity.

Section Review

SECTION: ELEMENTS

1. Sample answer: An element is an example of a pure substance.

2. Metals are shiny, are good conductors of heat and electric current, and are malleable and ductile. Nonmetals are poor conductors of heat and electric current, and, when they are solids, tend to be dull and brittle.

3. Sample answer: I would choose to make the container out of metal because metals are not brittle, so the container would not break if dropped. Also, metals are malleable, so forming the container out of this substance would be fairly easy.

4. Metalloids can be dull or shiny, may conduct heat and electric current, and are somewhat malleable and ductile. This list could be used to classify a substance as a metalloid if the substance has the properties of both a metal and a nonmetal. For example, if the substance is shiny, but also brittle, it could be classified as a metalloid.

5. $98.5\% - 46.6\% - 8.1\% - 5.0\% - 3.6\% - 2.8\% - 2.6\% - 2.1\% = 27.7\%$ silicon

6. Sample answer: I do not agree because the element could be a metalloid. Some metalloids are shiny.

SECTION: COMPOUNDS

1. A chemical change is needed to break down a compound.)

2. The three elements that make up table sugar are hydrogen, carbon, and oxygen.

3. $100.0\% - 51.5\%$ oxygen $- 42.1\%$ carbon $= 6.4\%$ hydrogen

4. When elements combine to form a compound, the compound's properties are different from the properties of the individual elements.

5. Sample answer: The jar does not contain a compound. The jar contains carbon and oxygen, but the two elements are not joined by chemical bonds.

SECTION: MIXTURES

1. solute

2. concentration

3. about 60°C

4. About 120 g more sodium chlorate than sodium chloride will dissolve.

5. Oxygen is the solvent and helium is the solute.

6. Carbon dioxide is a gas and is not very soluble in water. When the soda container is opened, the carbon dioxide will bubble out of solution, causing the soda to go flat. However, sugar is solid and is soluble in water, so the concentration of the sugar will not change.

7. Mixtures contain substances that are not chemically combined, so the properties of the components are the same as they would be outside of the mixture. The elements in a compound are chemically combined, and the properties of the elements are different from those of the compound. Mixtures do not have specific ratios of components, while compounds do have a specific ratio of elements. Mixtures can be separated by physical means, but the elements in a compound must be separated by chemical means.

8. Sample answer: I would use a magnet to separate the iron from the sawdust. The magnet will attract the iron but will not attract the sawdust.

Chapter Review

1. B
2. compound
3. element
4. nonmetal
5. solute
6. C
7. B
8. A
9. C
10. $100.0\% - 58.3\% - 4.2\% = 37.5\%$ carbon

11. % hydrogen in table sugar =
100% − 51.5% − 42.1% = 6.4%
The difference in percent hydrogen =
6.4% − 4.2% = 2.2%

12. Sample answer: I can tell that citric acid and table sugar are different compounds because they have different ratios of elements. A compound must have a specific ratio of elements.

13. An element cannot be separated into simpler substances. A compound can be separated chemically.

14. Nail polish is the solute, and acetone is the solvent.

15. An alloy is a solution, so 14-karat gold is not a pure substance. It is a mixture, not a compound.

16. The essay should develop the ideas that elements contain only one type of atom and cannot be broken down into simpler substances, that compounds are made when two or more different elements react to create a new substance, and that mixtures contain two or more substances that are not combined chemically. The essay should also have examples of elements, compounds, and mixtures.

17. An answer to this exercise can be found at the end of the Teacher Edition.

18. The powder is a compound. The change in color and the formation of a gas imply that a chemical change took place. Compounds can be broken down by chemical changes.

19. Sample answer: In mixtures, there is no chemical change in the original components. For example, if you took the grapes out of a fruit salad, they would still be grapes. Mixtures can be formed using any ratio of components. For example, you could make fruit salad by mixing 1 cup of grapes and 1 cup of melon. If you added another cup of melon, you would change the ratio of the components. But, the mixture would still be a fruit salad.

20. Carbonated beverages should be stored in a refrigerator. Gases are more soluble at lower temperatures, so more gas will stay dissolved in the beverage if it is kept cold.

21. Carbon monoxide and carbon dioxide are different because they have a different mass ratio of components. Carbon monoxide has a ratio of one oxygen atom to one carbon atom, and carbon dioxide has a ratio of two oxygen atoms to one carbon atom.

22. To form water, the atoms in hydrogen molecules and oxygen molecules must rearrange and combine so that two hydrogen atoms join with an oxygen atom.

23. (Teacher Note: The graph should have "dissolved solute" on the y-axis and "temperature" on the x-axis. The curve will decrease from left to right.) To increase solubility, you need to decrease the temperature. As the temperature decreases, more solute can dissolve.

24. Twice as much of the compound would dissolve, so 68 g would dissolve.

25. 50 g/200 mL = 0.25 g/mL

26. 150 mL × 0.6 g/mL = 90 g

27. Sample answer: To separate a mixture of salt, pepper, and pebbles, I would first pass the mixture through a screen that lets salt and pepper through but traps the pebbles. Because salt and pepper particles are about the same size, it should be easy to separate out the pebbles using this method. I would then mix the salt and pepper with water to dissolve the salt. Because salt is soluble, but pepper is not, I could collect the pepper by filtering the mixture. Finally, I would evaporate the water to recover the salt. Knowing the properties of matter helps me separate the substances because I know which substances can be separated by different physical means. For example, it is important to know that salt dissolves, so that I can separate that substance by filtering the mixture.

Section Quizzes

SECTION: ELEMENTS

1. B
2. E
3. A
4. D

5. C
6. B
7. D
8. C
9. D

SECTION: COMPOUNDS

1. A
2. C
3. D
4. B
5. A
6. D
7. B

SECTION: MIXTURES

1. E
2. H
3. B
4. C
5. I
6. J
7. F
8. D
9. G
10. A

Chapter Test A

1. B
2. B
3. B
4. C
5. B
6. C
7. B
8. A
9. A
10. C
11. D
12. B
13. E
14. F
15. A
16. D
17. B
18. C
19. G
20. dilute
21. distillation
22. nitrogen
23. nonmetals
24. ratio
25. concentration

Chapter Test B

1. B
2. C
3. B
4. A
5. C
6. A
7. B
8. A
9. A
10. D
11. A
12. B
13. C
14. A
15. B
16. C
17. D
18. E
19. A
20. B
21. F
22. D
23. D

Chapter Test C

1. compound
2. solute
3. alloys
4. pure substance
5. concentration
6. solution
7. A
8. B
9. D
10. C
11. A
12. Answers may vary. Sample answer: Compounds are considered pure substances because they are composed of only one type of atom.
13. Answers may vary. Sample answer: Metalloids, also called semimetals, are elements that have properties of both metals and nonmetals. Some metalloids are shiny, while others are dull. They are only somewhat malleable and ductile. Some metalloids, like silicon, are good electrical conductors only when mixed with other elements.

 Elements, Compounds, and Mixtures

14. Answers may vary. Sample answer: Salt water is a mixture, so its components can be separated by physical methods. Distillation involves heating the water, which changes water into steam. When salt water is distilled, steam is condensed back into water, and salt is left behind.

15. Answers may vary. Sample answer: I would first try to separate it by using distillation and a centrifuge. If I could separate it this way, I would know that it was a mixture. I would also try to separate it by passing an electric current through it. If I could separate it this way, I would know it was a compound. If I could not separate it at all, I might think it was an element, but I might also need to try other methods of separating it to be sure.

16. Answers may vary. Sample answer: Substances 1 and 3 are probably the same because they have the same melting point, the same amount of hardness, and are nonmagnetic. Substance 2 is different because it has a different density and hardness. It is also a slightly different color.

17. As pressure increases, the solubility of oxygen and nitrogen in water increases.

18. Oxygen experiences a greater change in solubility per unit pressure. The slope of the line that represents oxygen is steeper.

19. **a.** pure substances, **b.** compounds, **c.** solutions, **d.** physical or chemical, **e.** chemical

Performance-Based Assessment

3. 100°C

5. See chart below. These are sample data. Student data may vary.

Amount of Dissolved Salt

Cold water (mL)	Hot water (mL)
10	20

6. 104°C

8. The salt water boiled at a higher temperature than the plain water.

9. The hot water dissolved more salt.

10. Accept all reasonable answers. The first bubbles that form are caused by air that was dissolved in the water starting to escape. As the temperature increases, the water may start to evaporate, producing bubbles of water vapor that rise through the liquid.

11. Answers may vary. Sample answer: Hot water dissolves more solids than cold water dissolves.

12. Answers may vary. Sample answer: Water is a compound whose elements, hydrogen and oxygen, have been chemically combined to form a new substance with different properties than the original elements that made up the compound. Similarly, salt is made up of sodium and chlorine to form sodium chloride, or table salt. Salt water is a mixture because the salt and water form a solution. The salt dissolves in the water but does not chemically combine with it. Also, the salt can be separated from the water by the physical process of distillation.

Explore Activity

DATASHEET A

5. Sample answer: I used the color, the texture, and the hardness of the objects to classify them.

6. Size and shape are not reliable characteristics for classifying the objects because these two characteristics can change, but the objects themselves will remain the same.

7. **a.** Answers will depend on the objects that students examine.
b. Sample answer: If a substance is a solid at room temperature, then its melting point and boiling point must be greater than room temperature. If a substance is a liquid at room temperature, then its melting point is less than room temperature, but its boiling point is greater than room temperature. If the substance is a gas at room temperature, then both its melting point and its boiling point must be lower than room temperature.

DATASHEET B

5. Sample answer: I used the color, the texture, and the hardness of the objects to classify them.
6. Size and shape are not reliable characteristics for classifying the objects because these two characteristics can change, but the objects themselves will remain the same.
7. Answers will depend on the objects that students examine. If a substance is a solid at room temperature, then its melting point and boiling point must be greater than room temperature. If a substance is a liquid at room temperature, then its melting point is less than room temperature, but its boiling point is greater than room temperature. If a substance is a gas at room temperature, then both its melting point and its boiling point must be lower than room temperature.

DATASHEET C

5. Sample answer: I used the color, the texture, and the hardness of the objects to classify them because these are characteristic properties.
6. Size and shape are not reliable characteristics for classifying the objects because these two characteristics can change, but the objects themselves will remain the same.
7. Answers will depend on the objects that students examine. If a substance is a solid at room temperature, then its melting point and boiling point must be greater than room temperature. If a substance is a liquid at room temperature, then its melting point is less than room temperature, but its boiling point is greater than room temperature. If the substance is a gas at room temperature, then both its melting point and its boiling point must be lower than room temperature.

Quick Lab: Separating Elements

DATASHEET A

4. Yes. The iron nails were attracted to the magnet, but the aluminum nails were not.

5. Sample answer: A magnet can be used to separate materials that are magnetic from those that are nonmagnetic. In a recycling plant, a magnet could be used to separate cans containing iron from cans made entirely of aluminum.

DATASHEET B

4. Yes. The iron nails were attracted to the magnet, but the aluminum nails were not.
5. Sample answer: A magnet can be used to separate materials that are magnetic from those that are nonmagnetic. In a recycling plant, a magnet could be used to separate cans containing iron from cans made entirely of aluminum.

DATASHEET C

4. Yes. The iron nails were attracted to the magnet, but the aluminum nails were not.
5. Sample answer: A magnet can be used to separate materials that are magnetic from those that are nonmagnetic. In a recycling plant, a magnet could be used to separate cans containing iron from cans made entirely of aluminum.

Quick Lab: Identifying Compounds

DATASHEET A

5. Answers will vary depending on which compounds were used for A and B. The compound that fizzed was reacting with the vinegar, so that compound is the baking soda. The compound that did not react with the vinegar is powdered sugar.

DATASHEET B

5. The compound that fizzed was reacting with the vinegar, so that compound is baking soda. The compound that did not react with the vinegar is powdered sugar.

DATASHEET C

5. Baking soda will react with vinegar, but powdered sugar will not. The compound that fizzed was reacting with the vinegar, so that compound is the baking soda. The compound that did not react is powdered sugar.

Quick Lab: Identifying Solutes by Solubility

DATASHEET A

5. Answers will vary depending on which compound is used for A and for B. Students should explain that the substance that stopped dissolving is salt because salt is less soluble than sugar.

DATASHEET B

5. The substance that stopped dissolving is salt because salt is less soluble than sugar.

DATASHEET C

5. Answers may vary depending on which compound is used for A and for B. Students should explain that the substance that stopped dissolving is salt because salt is less soluble than sugar.
   Answers may vary. Sample answer: I could add about 8 times the amount of sugar I had added when the salt stopped dissolving.

Chapter Lab

DATASHEET A

9. Flame color is a test for the metal in each compound. Because each compound contains chloride, the color difference must be due to the different metals. Any color contribution from the chloride would be the same in each trial.

10. Answers will depend on the teacher's choice for the unknown compound. Students will know its identity because it will produce the same color flame as one of the other three test solutions.

11. The wire must be cleaned so that the color being observed is from the solution being tested, not from a mixture of two solutions.

12. Yes; sodium fluoride would likely burn the same color as sodium chloride because the flame test is a test for the metal in a compound, and both compounds contain sodium.

13. It is safe to use these compounds without a gas mask because the compounds combine chlorine with other elements and have different properties than chlorine gas.

14. Sodium metal reacts violently with water, and it would not be safe to test it using a flame test. However, the compound sodium chloride is safe to test because it is does not react in the same way as pure sodium.

DATASHEET B

9. Flame color is a test for the metal in each compound. Because each compound contains chloride, the color difference must be due to the different metals. Any color contribution from the chloride would be the same in each trial.

10. Answers will depend on the teacher's choice of the unknown. Students will know its identity because it will produce the same color flame as one of the other three test solutions.

11. The wire must be cleaned so that it is certain that the color being observed is from the solution being tested, not from a mixture of two solutions.

12. Yes; sodium fluoride would likely burn the same color as sodium chloride because the flame test is a test for the metal, and both compounds contain sodium.

13. It is safe to use these compounds without a gas mask because the compounds combine chlorine with other elements and have different properties than chlorine gas.

14. Sodium metal reacts violently with water, and it would not be safe to test it using a flame test. However, the compound sodium chloride is safe to test because it does not react in the same way as pure sodium.

DATASHEET C

9. The flame test is a test for the metal in each compound. Because each compound contains chloride, the color difference must be due to the different metals. Any color contribution from the chloride would be the same in each trial.

10. Answers will depend on the teacher's choice for the unknown compound. Students will know its identity because it will produce the same color flame as one of the other three test solutions.

11. The wire must be carefully cleaned so that it is certain that the color being observed is from the solution being tested, not from a mixture of two solutions.

12. Yes; sodium fluoride would likely burn the same color as sodium chloride because the flame test is a test for the metal, and both compounds contain sodium.

13. It is safe to use these compounds without a gas mask because the compounds combine chlorine with other elements and have different properties than chlorine gas

14. Sodium metal reacts violently with water, and it would not be safe to test it using a flame test. However, the compound sodium chloride is safe to test because it is does not react in the same way as the pure sodium.

Science Skills Activity

DATASHEET

1. The amount of Epsom salt and the amount of water are the variable parameters. I wanted to see how much Epsom salt would dissolve in varying amounts of water at a given temperature to determine solubility.

2. My controlled parameter is temperature. It is important to have controlled parameters in an experiment so that I can understand how the variable parameters change when all other conditions are kept the same. If too many parameters change at the same time, I will not be able to tell why my results change.

3. As the volume of water increased, the amount of Epsom salt that dissolved increased.

4. Answers may vary slightly, but the solubility should not depend on the volume of water. The solubility of Epsom salt at 20°C is 35.7 g/100 mL.

5. Answers may vary. Students may choose to use larger increments of Epsom salt, or less water, to speed up the experiment.